Mountains of fire

MOUNTAINS OF FIRE

The nature of volcanoes

ROBERT W. DECKER
and
BARBARA B. DECKER

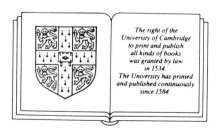

The right of the
University of Cambridge
to print and publish
all kinds of books
was granted by law
in 1534.
The University has printed
and published continuously
since 1584.

Cambridge University Press
Cambridge
New York   Port Chester   Melbourne   Sydney

Published by the Press Syndicate of the University of Cambridge
The Pitt Building, Trumpington Street, Cambridge CB2 1RP
40 West 20th Street, New York, NY 10011, USA
10 Stamford Road, Oakleigh, Melbourne 3166, Australia

© Cambridge University Press 1991

First Published 1991

Printed in the United States of America

Library of Congress Cataloging-in-Publication Data
Decker, Robert W. (Robert Wayne), 1927–
    Mountains of fire : the nature of volcanoes / Robert W. Decker and
Barbara B. Decker.
      p. cm.
    Includes index.
    ISBN 0-521-32176-X. – ISBN 0-521-31290-6 (pbk.)
    1. Volcanoes.   I. Decker, Barbara, 1929–   II. Title.
OE522.D37 1991
551.2′1 – dc20
                                        91-15691
                                        CIP

Contents

Color Plates are collected in a section facing page 102.

Preface

In the spring of 1982, a nearly unknown volcano in Mexico called El Chichón erupted for the first time in recorded history. Great explosions of ash produced massive, glowing avalanches of volcanic rock fragments that swept down the mountain's steep flanks. More than a thousand people were killed in villages on the mountainsides, and tens of thousands had to flee their homes.

Locally the El Chichón eruption was devastating, but its more subtle effects were felt around the world. Great quantities of dust and sulfur gases that had been lofted into the stratosophere by the explosions circled Earth for many months. Some scientists believe that one cause of the severe world weather of 1982–3 was this veil of volcanic dust and sulfuric acid aerosol from El Chichón. The stratosopheric haze was responsible, too, for the brilliant sunsets seen around the world for most of that year.

Volcanoes on Earth are only part of the story; in recent years, space probes have shown that volcanoes are some of the most pervasive features on other worlds. The Lunar Seas are great lava beds billions of years old. Olympus Mons on Mars is an ancient volcano 25 kilometers high and 800 kilometers wide. On Io, a moon of Jupiter, several active volcanoes are erupting giant spouts of material that appear to be molten or gaseous sulfur and sulfur compounds. Radar images through the dense sulfurous atmosphere of Venus, Earth's sister planet, show mountainous areas of probable volcanic origin.

Little is known about volcanoes on other planets and moons aside from the fact that they exist. To understand volcanoes on Mars we must first understand them on Earth. Fortunately, there are plenty of volcanoes to study; we know of more than 1,300 potentially active volcanoes on our planet, and about 50 of those erupt in an average year. The scientific knowledge gained from each eruption adds another piece to the intriguing puzzle of volcanism on Earth and around the universe. The purpose of this book is to shed some light on what volcanologists know, and what they hope to find out.

Active Volcanoes on the Earth's Surface

1. Hawaii
2. Alaska
3. Cascades
4. Mexico
5. Central America
6. Galapagos
7. Colombia and Ecuador
8. Peru and Bolivia
9. Chile
10. South Pacific
11. Iceland
12. Azores
13. Canary Islands
14. Cape Verde Islands
15. Cameroon
16. South Atlantic
17. Italy
18. Greece
19. Turkey

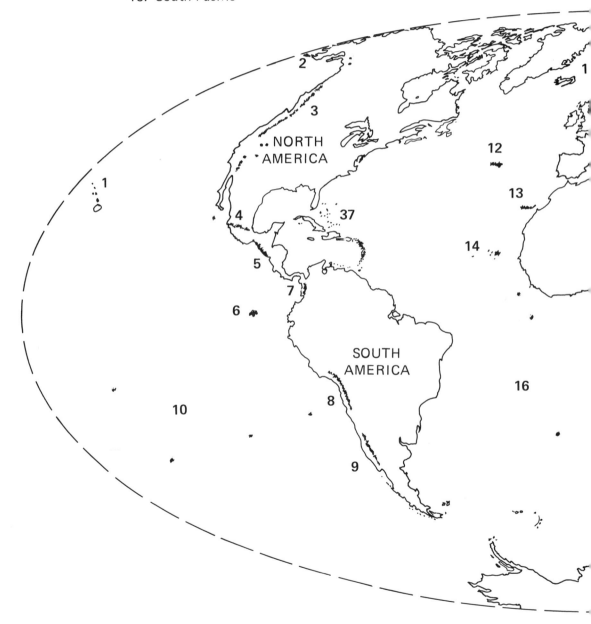

20. Iran
21. Middle East
22. East Africa
23. Indian Ocean
24. Kamchatka
25. Kurile Islands
26. Japan
27. Mariana Islands
28. Philippines
29. Indonesia
30. Papua New Guinea
31. Solomon Islands
32. New Hebrides
33. Samoa
34. Tonga Islands
35. Kermadec Islands
36. New Zealand
37. Caribbean

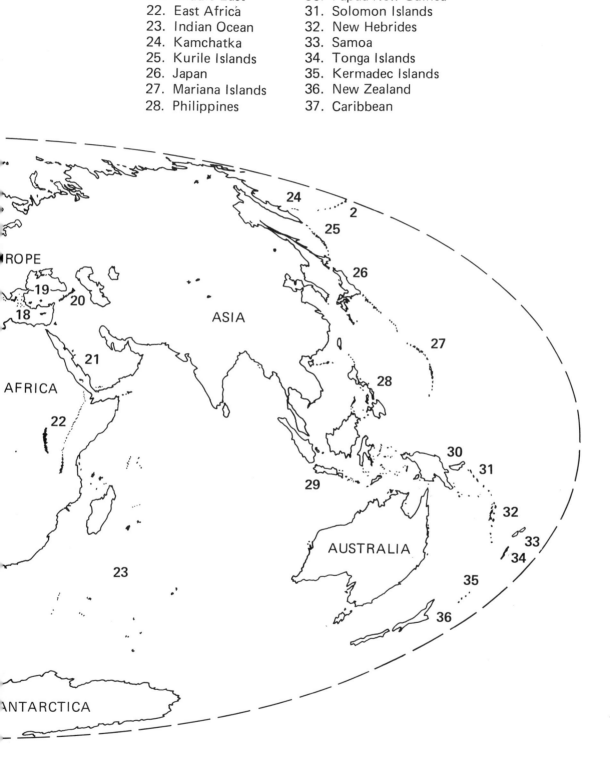

PART I

Volcanic mountains

Men argue; Nature acts.
– Voltaire

I

What is a volcano?

It was past 11 o'clock at night; most of the 25,000 people living in Armero were in bed, but many were too frightened to sleep. They had heard ominous rumblings from the towering, ice-capped Ruiz Volcano earlier in the evening, and a light ashfall had blanketed the town. Then, as one survivor described it, "The world screamed."

The ground shook and the roar was so loud that people shouting at close range could not hear one another as layer upon layer of mud, rocks, and debris with the consistency of wet concrete trapped, drowned, and covered 22,000 men, women, and children. Thick waves of hot mud washed down the Lagunillas River Valley and sulfur fumes filled the air. Armero was buried alive (Fig. 1.1). Hell could not have been more frightening. Only those who had evacuated when the ashfall occurred, those who lived on higher ground above the river, and the lucky few who were swept to the thinner edges of the mudflows survived.

The volcano has a history of previous eruptions, notably in 1595 and early in the nineteenth century. The ancient Indian name of the mountain was Cumanday, "the smoking nose." In 1845 a volcanic eruption or earthquake sent mudflows pouring down the Lagunillas River, killing about 1,000 people. According to scientists who have studied the November 13, 1985, eruption of Nevado del Ruiz, a complex sequence of events led to the catastrophe at Armero. Ruiz is a broad, nearly flat-topped volcano capped by an extensive snow-and-ice field. At 5,389 meters, it is one of the highest peaks in Colombia. It lies along the crest of a sharp Andean mountain ridge that separates two deep north–south trending valleys.

Before the 1985 eruption, vents spewing steam and other volcanic gases had been continuously active in a crater near the northern rim of the ice sheet, and a small acid lake filled the crater bottom. An earthquake

Figure 1.1. A mudflow generated by an eruption of Nevado del Ruiz Volcano in Colombia on November 13, 1985, devastated the town of Armero and killed 22,000 people. The business district was entirely swept away; only the vague outlines of foundations show through the drying mud. Houses on the higher ground to the right were partially destroyed. (Photograph taken December 9, 1985, by Richard Janda, U.S. Geological Survey)

swarm beneath Ruiz began in November 1984, one year before the eruption. This probably marked the beginning of a new rise of molten rock from greater depth to a zone a few kilometers below the surface.

Steam emission increased in December 1984 and a small explosion occurred on September 11, 1985, but the ashes and rocks from this eruption were old volcanic material from the crater region; no new molten rock, called "magma" when underground, had yet reached the surface. Heat and gases from the shallow intrusion of magma, however, caused the fuming gas vents to increase in activity and formed new deposits of sulfur on the summit of Ruiz (Fig. 1.2).

The tragic eruption of November 13 began about 3:00 P.M., with small explosions from the crater and minor ash falls to the northeast. Just after 9:00 P.M. the activity became more violent, and new magma reached the surface in a jetting column of hot pumice fragments and ash. The timing, distribution, and size of the fallout all indicate that the main

Figure 1.2. Small plumes of steam and other volcanic gases issued from the crater of Nevado del Ruiz before the November 13, 1985, eruption. Although the 5,389-meter-high mountain is only 5 degrees north of the equator, its summit is covered by a large snow-and-ice cap. (Photograph taken September 25, 1985, by Marta Calvache, Observatorio Vulcanologico de Colombia)

eruption lasted only about two to three hours, and that it formed a moderately high ash cloud.

Although this was a relatively small eruption, the hot blanket of pumice and ash melted large quantities of water from the snow-and-ice fields on the summit of Nevado del Ruiz. As the melt-water flood rushed down the valleys on the steep north flanks of Ruiz it picked up soil, rocks, and trees and churned itself into devastating mudflows (Fig. 1.3). Debris-laden mudflows sped down the steep, narrow Lagunillas canyon that drops more than 4,000 meters on its twisting 60-kilometer course to the Magdalena Valley, and spread out over wide areas as they swept beyond the confines of Lagunillas canyon. Unfortunately, Armero lay right at the canyon's mouth.

The mudflows are estimated to have raced down the mountain at speeds of 35 kilometers per hour, stripping off trees and soil to heights of 80 meters above the canyon's normal river level. At the canyon mouth where the flow spilled out on Armero the waves of mud were as high as 30 meters, but these quickly spread out to floods about 4 meters deep.

Two or three major waves of mud swept through in 20 to 30 minutes;

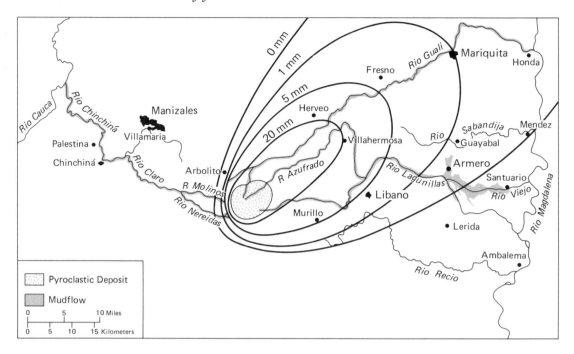

Figure 1.3. Map of Nevado del Ruiz Volcano and vicinity showing deposits of the November 13, 1985, eruption. Contours in millimeters indicate the thickness of the ashfall. Less than 10% of the ice cap was melted by the pyroclastic (hot fragmental) deposit. (From Barry Voight, Pennsylvania State University; data from the Observatorio Vulcanologico de Colombia)

the early mud was cold, but some survivors described the later waves as hot. At its maximum flow, an estimated 47,000 cubic meters per second of mud and debris-laden water swept down the Lagunillas River, a rate equal to one-fifth of the flow of the Amazon River pouring down a narrow canyon. The deposit of mud left behind after the water flowed away was about 1.0 to 1.5 meters thick over an area of 33 square kilometers, but boulders up to 10 meters in diameter that lie scattered in the muck testify to the violence of Armero's awful destruction.

Ninety percent of the original snow-and-ice field still remains on Nevado del Ruiz; continuing earthquakes and small eruptions indicate that the volcano's present activity is not over. The survivors of Armero and the residents of other towns at the foot of Ruiz fearfully await the volcano's next move.

WHAT IS A VOLCANO?

Erupting volcanoes are mountains gone mad, as the residents of Armero found to their sorrow. Accounts of their violent behavior appear in the mythology of almost all cultures that evolved in volcano country: Greeks, Romans, Indonesians, Japanese, Icelanders, and Hawaiians all had gods or goddesses of fiery volcanoes. Generally the beauty and serenity of those deities prevailed, but when aroused and angry they were poised ready to rain destruction on the countryside.

The word *volcano* comes from the island called Vulcano, off the southwest coast of Italy. Because of frequent eruptions on that island, Romans considered it to be the forge of Vulcan, the god of fire and maker of weapons. The word has since come to mean any vent in the Earth's crust through which magma reaches the surface.

The term *volcano* is also used to describe the landform built by eruptions from the vent. Just which meaning is implied depends on the usage of the word. For example, saying that Mauna Loa Volcano spewed lava over many square kilometers of land refers to the volcano as a vent. On the other hand, to say that Mauna Loa is the largest volcano on Earth refers to the volcano as a mountain.

VOLCANOES, DEAD OR ALIVE

Because volcanic activity is almost as diverse as human nature, volcanologists use some humanlike terms to describe a volcano's various moods: *alive, active, restless, awakening, erupting, in repose, dormant, sleeping, dead,* and *extinct* are words often used to characterize a volcano's various states. Considering that the life-span of a volcano may exceed a million years, some caution must be taken in using these terms. To clarify the meanings in regard to volcanic activity, it is best to use examples.

In the 22-volume *Catalogue of Active Volcanoes of the World,* the International Association of Volcanology defines an *active* volcano as one that has erupted during historic time. By this definition there are more than 500 active volcanoes in the world. The only problem with this classification is that the span of recorded history differs greatly from region to region. In the Mediterranean, recorded history reaches back nearly 3,000 years; in Hawaii, only 200 years; and on some uninhabited Aleutian Islands in Alaska the records still have gaps. Not only is written history erratic in length and detail, even the oldest records cover only a small fraction of a volcano's geologic lifetime.

The earliest known record of a volcanic eruption is a wall painting at Catal Huyuk, an ancient village in central Turkey. James Mellaart, the archaeologist who excavated the site, believes the painting depicts an eruption of Hasen Dag Volcano in about 6200 B.C., a nearby volcano with no written record of eruption. In written history, the first eyewitness account of a major volcanic eruption dates from A.D. 79, when a Roman, Pliny the Younger, vividly described in a letter the huge eruption of Vesuvius that buried Pompeii and Herculaneum, killing his uncle, Pliny the Elder.

The Smithsonian Institution in Washington, D.C., has attempted to solve the limitation of recorded history by including in their list of world volcanoes all volcanoes that have apparently erupted in the past 10,000 years – the period of time since the last major world glaciation. The Smithsonian list tallies 1,343 potentially active volcanoes worldwide.

Both the International Association of Volcanology and the Smithsonian's volcano catalogs are working lists. If an eruption occurs at a volcano that has had no known eruptions – the 1982 eruption of El Chichón in Mexico, for example – that volcano is added to both lists. Similarly, if evidence is found that indicates a prehistoric eruption of a volcano not included on the Smithsonian's list, that one will be added to their tally.

An *erupting* volcano is one emitting molten lava or ejecting solid fragments of volcanic material such as ashes, cinders, or blocks. A volcano exhaling gases such as steam, carbon dioxide, or sulfur gases, either from small vents or as a plume of gases from its crater, is listed as a potentially active volcano but is not considered to be in eruption. Volcanoes giving off gases but not erupting are referred to as being in a "fumerolic stage."

A potentially active, or *live,* volcano is one that will probably erupt in the future; an *extinct,* or dead, volcano is not likely to erupt again. A journalist thinking in human life-spans may tend to consider a volcano extinct if it has not erupted in historic time; a geologist, however, would prefer a time span on the order of hundreds of thousands of years without an eruption before putting a volcano in the "probably extinct" category. This vast difference in scale between time in human terms and geologic time is at the root of many misunderstandings and misconceptions about volcanoes and other geologic phenomena.

A *dormant,* or sleeping, volcano is a live volcano that is not currently erupting. The period between eruptions is called the volcano's *repose time.* On Mauna Loa Volcano the average repose time is about 3 to 4 years, but since 1832, when record keeping for that volcano began, Mauna Loa's actual repose times have been as short as a few months and as long as 25 years. Neighboring Mauna Kea Volcano, with several astronomical observatories on its summit, has been dormant for about 4,000 years. However, considering Mauna Kea's life-span of about 500,000 to 1 million years, a 4,000-year slumber is not long enough to

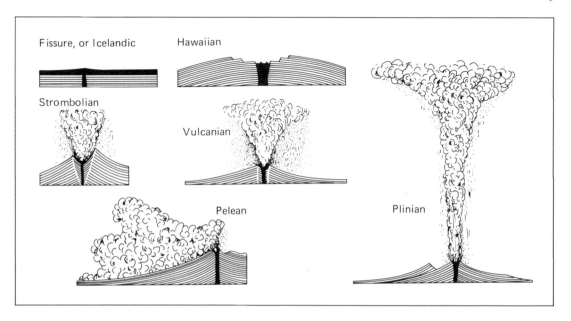

Figure 1.4. Schematic cross sections of major types of volcanic eruptions. (After Arthur Holmes, Principles of Physical Geology, *2nd ed. [New York: 1965], Ronald Press, p. 305)*

consider it dead. Even so, the great telescopes with useful life-spans of perhaps 100 years will probably have served their purpose long before Mauna Kea awakens again.

TYPES OF VOLCANOES

Volcanoes are classified both by form and by eruptive habits. In the nineteenth century, geologists proposed that eruptions be classed by types of volcanoes that had characteristic habits. This scheme caught on and has been modified to a list of six types. In increasing degrees of explosiveness, these are: Icelandic, Hawaiian, Strombolian, Vulcanian, Peléan, and Plinian (Fig. 1.4).

Icelandic volcanic eruptions generally (but not always) involve out-pourings of hot, relatively fluid molten lava from lengthy fissures that are sometimes as long as 25 kilometers. Gases dissolved in the erupting magma boil out at surface pressures, forming spectacular lava fountains along the erupting fissures. The recent eruptions of Krafla Volcano in northern Iceland are examples of this type.

These fluid, Icelandic-type eruptions build a plateau composed of thick, nearly flat layers of hardened lava that has flowed many tens of kilometers from the vents. The vents themselves may be marked by a row of low hills built up by hot, solid cinders or molten spatter ejected from the fissure. However, subsequent lava flows from parallel fissures are apt to bury these hills during later eruptions. Other volcanic land-forms occur in Iceland, but much of the island is made up of lava plateaus that were formed by fissure eruptions.

Hawaiian volcanoes also erupt from fissures at the mountains' summits and along weak zones on their sides called "rift zones." They are similar in many ways to Icelandic eruptions, with high lava fountains and long, fluid flows fed by rivers of incandescent lava. Because summit eruptions occur more frequently than do rift eruptions, however, the tops of Ha-waiian volcanoes grow upward faster than their flanks, and form gently sloping domal mountains rather than lava plateaus (Fig. 1.5). These great piles of hardened lava, like giant mounds of candle drippings, are called "shield volcanoes," a name first used in Iceland, because their shape resembles a warrior's shield lying faceup on the ground. The numerous eruptions of Kilauea Volcano are good examples of the Hawaiian type.

Strombolian eruptions take their name from Stromboli, a small volcanic island off the southwest coast of Italy that has been in almost constant eruption for centuries. Almost every 15 to 20 minutes a small explosion of bursting gas throws a shower of clots of pasty, incandescent lava into the air. The lava surface in the summit crater then crusts over until accumulating gases burst forth again. The regularity of these small ex-plosive eruptions has earned Stromboli Volcano the title "Lighthouse of the Mediterranean."

Stromboli is a steep, conical mountain. Its shape is largely determined by the fact that it erupts from a more pipelike conduit rather than from a fissure. The molten rock that spatters out from the crater at its summit, and the thick lava flows that mark its larger eruptions, are too viscous to flow far from their vent, so a steep cone develops.

Stronger explosive eruptions that form dark ash clouds mainly com-posed of steam, other volcanic gases, and solid fragmental material are called *Vulcanian* eruptions after Stromboli's neighbor island Vulcano. The ash clouds from these eruptions rapidly expand into many turbulent lobes and are often described as cauliflower-shaped. After the initial explosive phase, Vulcanian eruptions also produce thick, sluggish lava flows. In each period of eruption, the ash deposits generally precede the extrusion of viscous lava flows, and the alternating layers of ash and lava build a steep cone called a *stratovolcano* or *composite cone*.

After an eruption of Mont Pelée Volcano in 1902 killed the 29,000 inhabitants of the Caribbean port of Saint Pierre, geologists added *Peléan* to the list of eruption types. Peléan eruptions are extremely destructive

Figure 1.5. Mauna Loa, a gently sloping shield volcano in Hawaii, rises to a height of 4,169 meters above sea level–10,000 meters above the sea bottom. This aerial view toward the northeast was taken just west of South Point; the summit of Mauna Loa is 70 km distant. The dark young lava flows have erupted from Mauna Loa's southwest rift. (Photograph by the U.S. Geological Survey)

because they generate high-speed avalanches of hot ashes that have been mobilized by expanding gases. These *nuées ardentes* (glowing clouds) are so charged with volcanic ash that they are heavier than air, and pour down the steep flanks of an erupting volcano at speeds in excess of 100 kilometers per hour. The dense, lower part of these hot mixtures of volcanic fragments and gas are also called "glowing avalanches," "ash flows", or "pyroclastic flows." Whatever the name, an ash flow is a major danger in a volcanic eruption. It cannot be outrun, and it annihilates everything in its path because of its high temperature and destructive force.

Ash flows and mudflows often build an apron of deposits surrounding the steeper stratovolcano that is their source. This combination of land-forms and volcanic products indicates that eruptions of more than one type have occurred at that explosive volcano.

Plinian is the term used for extremely explosive eruptions similar to the eruption of Vesuvius in A.D. 79. A fundamental characteristic of this type of eruption is the sustained jetting of volcanic ash into a high cloud. Thick blankets of ash and pumice fall to the ground downwind, while fine ash and volcanic gases are injected into the stratosphere, where they may affect weather and climate. Large ash flows are also produced during Plinian eruptions.

The sustained jetting is often so violent that it rips away large portions of the summit crater. Progressive downward exposure of the magma column can sustain these ash eruptions for hours. A Plinian eruption is not a single great explosion; it is more analogous to a burning rocket engine buried in the ground with its nozzle pointed upward. However, in a Plinean eruption there is no combustion.

The energy of a Plinian eruption comes from explosive boiling of volcanic gases dissolved in the erupting magma. This type of eruption is sustained until the magma column is cored out so deeply that the upper parts of the volcano collapse in a roughly circular basin and choke off any further explosive release of gas. Eruptions of enormous volume sometimes remove so much magma from beneath the volcano that col-lapse basins engulf the summit. These great circular basins at the summits of volcanoes are called *calderas*. They are often as wide as several kilo-meters in diameter and hundreds of meters deep.

One shortcoming of the use of eruption types to classify volcanoes is that some characteristic landforms – such as calderas and cinder cones – can form at several different types of volcanoes. Calderas form by collapse in Iceland and Hawaii as well as at major explosive volcanoes. Cinder cones, sometimes called "scoria cones," can form from the fallout of the cooled fragments from persistent lava fountains in Iceland and Hawaii, as well as from Vulcanian eruptions (Fig. 1.6).

Like all generalities, classification of eruption types and volcanic land-

Figure 1.6. Cerro Negro Volcano in 1968. This 230-meter-high cinder cone in Nicaragua has been erupting intermittently since its birth in 1850. Notice how the cinder cone covers older lava flows in the foreground. New lava flows issue from a vent on the left base of the cone. (Photograph by U.S. Geological Survey)

Figure 1.7. Shishaldin in the Aleutian Islands, Alaska, is a beautifully symmetrical strato-volcano. Since 1775, about 30 small to moderate explosive eruptions from this 2,857-meter-high peak have been observed, the latest in 1987. The small plume of volcanic gases seen in this photograph is not considered an eruption. (Photograph by U.S. Geological Survey)

forms can be taken only so far. Volcanic eruptions are caused by many factors, which are explained in more detail in Chapter 3, and the shape of a volcanic mountain may be the end result of a series of eruptions that vary in location, type, size, and sequence. It takes years of detailed geologic mapping to unravel the complex eruptive history of an individual volcano.

Nevertheless, it is generally true that the dangers inherent in volcanic eruptions increase with their explosiveness. Icelandic and Hawaiian eruptions usually involve relatively "quiet" emissions of lava, which often destroy property but seldom kill people. Peléan and Plinian ash-flow eruptions are extremely lethal, often killing thousands of people.

Volcanic landforms also offer clues to the potential danger of future eruptions. Lava plateaus and shield volcanoes are formed by relatively

Figure 1.8. Crater Lake in Oregon fills a 10-km-wide caldera. The lake's surface elevation is 1,882 meters and it reaches a depth of 589 meters. The caldera formed 6,900 years ago as 50 cubic km of magma were explosively erupted and the ancient volcano's summit collapsed into the emptied portion of the underlying magma chamber. Wizard Island, a cinder cone, grew after the great caldera-forming eruption. (Photograph by the Washington State National Guard)

quiet eruptions of fluid lavas. In contrast, stratovolcanoes are largely built of deposits from explosive eruptions (Fig. 1.7). Vesuvius and Mont Pelée are stratovolcanoes notorious for their destructive eruptions. Calderas formed by collapse when huge volumes of ash-flow deposits are violently expelled are evidence of even more catastrophic eruptions (Fig. 1.8). Fortunately, caldera-forming eruptions are rare events.

Naming the types of volcanoes, while helpful, is only a beginning in trying to understand them. Some answers to such questions as where volcanoes occur and why they erupt lie ahead. The next chapter examines the physical setting of volcanoes, and explains why they are not scattered randomly about the Earth but usually occur along great mountain chains, both on land and beneath the sea.

2

Volcanic belts

Mount St. Helens, Washington: March 20, 1980

After 123 years of repose, Mount St. Helens shook itself awake with a swarm of earthquakes. Frequent shakes were still being felt by local residents when, a week later, small ash eruptions began to excavate a new crater in the summit glacier. Ash from hundreds of these small explosions covered the white snow-capped mountain with a black shroud during the next several weeks.

As the earthquakes continued, an ominous bulge high on the north flank of Mount St. Helens grew at the rate of 1 to 2 meters a day, a change visible to the naked eye. Most geologists studying the activity concluded that an intrusion of magma was being injected at a shallow depth beneath the north side of Mount St. Helens, and that this forceful injection of molten rock was causing the earthquakes and the bulge. The main questions were whether — and when — this new magma might reach the surface.

The answers came on May 18, 1980. At 8:32 A.M., a sharp earthquake dislodged a huge avalanche of rock and ice from the north face of Mount St. Helens that had been oversteepened by the growing bulge (Fig. 2.1).

This removal of an enormous mass of rock and ice suddenly released the pressure on the superheated ground water and shallow magma beneath the volcano, like the popping of a lid on an enormous pressure cooker. The resulting blast of explosively expanding steam and volcanic gases tore off the remaining north side of the peak, ground it to fragments, and hurled it across 550 square kilometers of forested ridges. Fifty-seven people in the area were killed; two of them had radios and transmitted stark last words.

David Johnston, a scientist with the U.S. Geological Survey, was making measurements of the bulge from a high ridge 9 kilometers to

Figure 2.1. Drawings from eyewitness accounts and published photographs of the first seconds of the major eruption of Mount St. Helens on May 18, 1980. Top: *Bulge is visible on north face before eruption.* Center: *Huge avalanche of entire north side of the cone.* Bottom: *Twenty seconds later the giant steam eruption begins to override the avalanche. The sudden release of pressure within the volcanic cone as the avalanche slid away allowed the heated ground water and gases dissolved in the shallow intrusion of magma to boil out explosively. (Drawings by Richard Hazlett)*

the north of Mount St. Helens. As the eruption began, Dave urgently radioed to his headquarters "Vancouver, Vancouver – this is it – " just before the explosion cloud engulfed him. Gerald Martin, a retired navy radio operator, was manning the Washington Department of Emergency Services volunteer warning station 3 kilometers north of David Johnston's ridge. Martin coolly reported the avalanche and the start of the blast. About a minute later his last transmission was "The camper and the car just over to the south of me [David Johnston's camp] are covered. . . . It is going to get me, too."

Other witnesses were more fortunate. Geologists Keith and Dorothy Stoffel were flying in a light plane just over the summit when the avalanche began. In their words:

> As we approached the summit, flying at an altitude of 11,000 feet [Mount St. Helens was 9,677 feet – 2,950 meters – high at that time] everything was calm. Just as we passed above the western side of the summit crater, we noticed landsliding of rock and ice debris inward into the crater. The pilot tipped the wing towards the crater, giving us a better view of the landsliding. The north facing wall of the south side of the main crater was especially active. Within a matter of seconds – perhaps 15 seconds – the whole north side of the summit crater began to move instantaneously. As we were looking directly down on the summit crater, everything north of a line drawn east–west across the northern side of the summit crater began to move as one gigantic mass. The nature of the movement was eerie, like nothing we had ever seen before. The entire mass began to ripple and churn up, without moving laterally. Then the entire north side of the summit began sliding north along a deep-seated slide plane. We were amazed and excited with the realization that we were watching this landslide of unbelievable proportions slide down the north side of the mountain towards Spirit Lake. We took photographs of this slide sequence occurring, but before we could snap off more than a few pictures, a huge explosion blasted out of the landslide-detachment plane. We neither felt nor heard a thing, even though we were just east of the summit at that time.
>
> From our viewpoint, the initial cloud appeared to mushroom laterally to the north and plunge down. Within seconds, the cloud had mushroomed enough to obscure our view. At about this time, the realization of the enormous size of the eruption hit us, and we focused our attention on getting out of there. The pilot opened full throttle and dove quickly to gain speed. He estimated that we were going 200 knots [370 kilometers per hour]. The cloud behind us mushroomed to unbelievable dimensions and appeared to be catching up with us. Since the clouds were billowing primarily in a northerly direction we

turned south, heading straight toward Mount Hood. [This turn saved their lives; if they had continued due east or turned north, the blast cloud, expanding east, north, and northwest at 400 kilometers per hour, would have outraced their aircraft.] To the east of the volcano the ash cloud separated into billowing mushroom-shaped clouds and a higher overhang of cirrus-type clouds. Ashfall from the mushroom-shaped clouds was heavy. Lightning bolts shooting through the clouds were tens of thousands of feet high. Soon the ash extended to altitudes higher than 50,000 feet [15,240 meters]. We thought about flying back to Yakima . . . but decided against it, realizing we could never beat the ash cloud. Sometime between 9:00 and 9:15 A.M. we landed at Portland airport.

The gases kept boiling out of the exposed magma body for the next nine hours, jetting ashes into the high cloud and sending ash flows down the breached north flank of Mount St. Helens. Floods of mud poured down the streams draining the mountain, and fine volcanic ash in deposits as much as 4 centimeters thick fell for hundreds of kilometers to the east. By evening the main eruption was over, and the ugly stump of Mount St. Helens was only 8,364 feet high – a loss of 1,313 feet (400 meters).

VOLCANOES AND PLATE TECTONICS

Most volcanoes occur in belts or chains, like great beads on a giant string. Sometimes the chains are quite straight, but more often they form gently curving arcs. The "Ring of Fire" around the Pacific Ocean is a series of great arcs of volcanoes, which together form a nearly complete circle around the water hemisphere of the Earth (Fig. 2.2).

One segment of the Ring of Fire extends from northern California's Lassen Peak in the south, for a thousand kilometers with a dozen volcanic peaks to British Columbia's Mount Garibaldi in the north. These are the Cascade Volcanoes, and Mount St. Helens is one of them.

In addition to the circum-Pacific ring of volcanoes, another major volcanic belt extends from the Mediterranean Sea through Iran and, after a gap, continues through Indonesia to the Pacific belt. Indonesia has 127 live volcanoes, second only to the United States as the country on Earth with the largest number of potentially active volcanoes.

The Ring of Fire and the Mediterranean–Indonesian belt together account for 80 percent of the world's known volcanoes on land. However, a conventional world map tells only part of the story; what it does

Figure 2.2. Earth topography from digital elevation data shows the three major settings of volcanoes. Subduction volcanoes occur in the island arcs and mountain ranges that surround the Pacific basin. Rift volcanoes occur along the crests of the Mid-Atlantic Ridge and other mid-oceanic ridges. The Hawaiian hot-spot volcanoes are marked by the narrow dog-leg ridge in the north-central Pacific. (Computer generated image by Dr. Peter W. Sloss, NOAA, National Geophysical Data Center, Boulder, Colorado)

not reveal is that most of the world's volcanoes are hidden beneath the oceans. Maps of the terrain under the sea show great mid-ocean ridges that are apparently capped by hundreds of live volcanoes.

For many years geologists and geographers puzzled as to why most volcanoes occur in linear or arcuate chains rather than at random points on the Earth's surface; now the concept of plate tectonics offers an answer to this question. The idea that segments of the crust of the Earth are slowly moving over large horizontal distances – thousands of kilometers – was first seriously proposed in the early 1900s by Austrian scientist Alfred Wegener. He called his concept "continental drift" and envisioned rafts of rigid continental crust slowly drifting through a highly viscous oceanic crust, an idea that most scientists of his day would not accept.

Wegener's idea was given new life and a new twist in the 1960s when oceanographers and geophysicists discovered that the seafloor crust is made up of belts of rock that are magnetized in long, linear patterns, parallel to the mid-ocean ridges and symmetrical on both sides of the ridges. These patterns indicate that new seafloor is being created at the mid-ocean ridges and slowly spreads away from the rifts in opposite directions at speeds of 1 to 10 centimeters per year: about the rate at

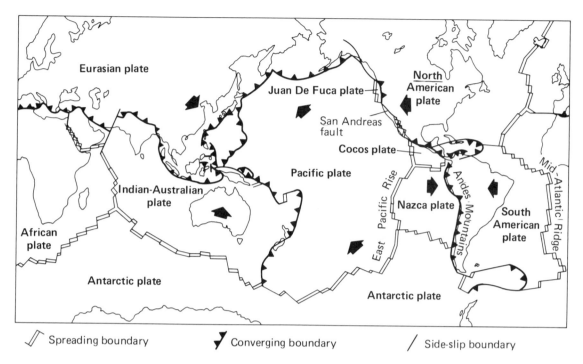

Figure 2.3. The mobile shell of the Earth is broken into tectonic plates. The spreading edges are rift zones and the converging edges are subduction zones. Volcanoes are common on both these boundaries. (After Warren Hamilton, U.S. Geological Survey)

which fingernails grow. The oceanographers called this process "seafloor spreading," and the geophysicists called it "plate tectonics."

According to the plate tectonics concept, the Earth's surface is broken into a dozen major plates 50 to 100 kilometers thick, which move about horizontally with respect to one another (Fig. 2.3). These plates "float" and "slide" on a highly viscous layer in the Earth's mantle beneath them. In comparison to the Earth's diameter the plates are comparable to broken pieces of shell on a soft-boiled egg. The major difference from Wegener's idea is that many of these plates are made up of both continental and oceanic crust, with the continents riding passively on the spreading seafloor like ships locked into a drifting ice floe.

The main action in plate tectonics occurs at the edges of the plates – earthquakes and volcanic eruptions on human time scales; splitting, shifting and crumpling continents on geologic time scales. Between separating plates is a zone of "healed" cracks, such as the Mid-Atlantic Ridge; converging plates override one another by thrusting and folding rocks into great mountain ranges like the Himalayas; and sideslipping plates create notorious fault zones like the San Andreas in California, the Alpine in New Zealand, and the Anatolia in Turkey.

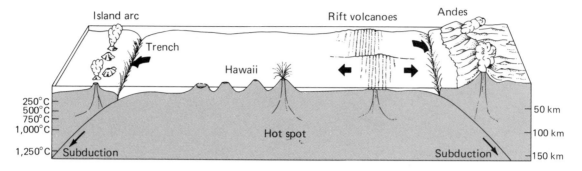

Figure 2.4. The major types of volcanoes related to plate tectonics. Subduction zone volcanoes occur on the overriding plate near converging boundaries; rift volcanoes—mainly submarine—occur along the spreading boundaries; and hot spots beneath the plates form alignments of volcanoes that become extinct "downstream" from the hot spot. (Modified from Les Observatoires Volcanologiques Français, Institut de Physique du Globe de Paris *[1989], p. 15)*

RIFT VOLCANOES

Where tectonic plates are moving apart, volcanoes fill the separating scars of the diverging plates with lava flows, creating new seafloor along the crests of mid-ocean ridges (Fig. 2.4). Such volcanoes are known as rift volcanoes, with the best-known on-land examples in Iceland and in the rift valley of East Africa.

Only a small fraction of the 70,000-kilometer-long world rift system is above sea level, but in these areas there are about 250 live rift volcanoes. It is inferred that many more unknown but potentially active volcanoes, perhaps several thousand, have erupted quietly in deep water along the mid-ocean ridges during the past 10,000 years.

On average, fewer than one rift volcano on land erupts each year, but perhaps 20 hidden rift eruptions occur beneath the sea on a yearly basis. Not one of these eruptions has ever been observed, although earthquake swarms that have been located by distant seismographs give some hint of their position. Observing and studying rift-volcano eruptions along mid-ocean ridges at depths of 2,000 to 3,000 meters beneath the sea is one of the great challenges in volcanology today.

SUBDUCTION VOLCANOES

Volcanoes also build high peaks along the crumpled zones caused by the slow collision of converging plates. In this situation, the volcanoes do not form at the exact contact of the plates; that zone is generally a deep oceanic trench. Plates converge by one overriding the other, usually with an oceanic edge plunging or subducting beneath a continental edge. The portions of plates covered by deep oceans are denser and thinner than the parts of plates that carry continental land masses.

Subduction-related volcanoes occur about 200 kilometers inland from the oceanic trench. The reason they are so far inland from the plate margin is not clear, but it is probably related to a zone of extensive vertical fractures caused by the buckling of the overlying plate. Also, these volcanoes are over the general area where the subducting plate has reached a depth of about 100 kilometers; the high temperatures within the Earth at this depth, and the lowered melting temperatures of the rocks (caused by the addition of water and carbon dioxide from seafloor sediments that are dragged down on the subducting plate) seem to make this an ideal region for the formation of large batches of magma.

The Ring of Fire volcanoes around the rim of the Pacific Ocean, and the Mediterranean–Indonesian volcanic belt, all belong to the subduction-volcano clan. Because the subducting plates generally push beneath the overriding plates at angles of 30 to 60 degrees, these contact zones trace arcs on the sphere of the Earth. The island arcs of New Zealand, the southwest Pacific, Indonesia, the Philippines, Japan, Kuriles, Kamchatka, the Aleutians, and the Caribbean are famous for their subduction volcanoes. The western spine of North and South America – the Cascade Range, the highlands of Mexico and Central America, and the Andes Range – complete the eastern side of the Ring of Fire.

About 1,000 live subduction volcanoes occur along the edges of converging plates, and in any one year about 40 of these will be in some state of eruption. It is clear from location alone that volcanic activity and plate tectonics are closely related; the edges of the tectonic plates account for more than 95 percent of the world's active volcanoes.

HOT-SPOT VOLCANOES

Hawaiian volcanoes, occurring well within the Pacific Plate and 4,000 kilometers from the nearest plate edge, are the best-known exceptions to the rule that most volcanoes are found at plate margins. To explain their anomalous position and yet relate them to plate tectonics, Canadian geophysicist J. Tuzo Wilson developed the hot-spot idea.

In this concept, persistent zones of magma supply occur deep within the Earth at several places beneath the moving plates. These hot spots produce large batches of magma that rise through the overlying plate. The magma rises buoyantly because it is lighter than its surrounding rocks, and earthquake patterns suggest that it wedges apart cracks to pierce the plate. Through geologic time as the plate moves over a hot spot, new volcanoes appear as the older, extinct volcanoes are carried away in the direction of plate motion. An analogy would be smoke signals drifting away in a gentle wind.

Wilson's hypothesis explained what geologists had puzzled over for more than a century – why the Hawaiian Islands have active volcanoes only on the southeast end of the chain, and why the islands become progressively older to the northwest. Moreover, knowing the approximate rate of motion of the Pacific Plate, his hot-spot concept predicts what the age of the older volcanic islands should be. Dating of the various Hawaiian Islands by measuring the minor amounts of radioactive elements and their daughter products in ancient Hawaiian lava flows indicates progressive increase in age to the northwest consistent with the hot-spot concept.

Kauai is 5 million years older than, and 500 kilometers northwest of, the Big Island of Hawaii. The overall track of the Hawaiian Hot Spot – marked by the submarine Hawaiian Ridge and Emperor Seamounts – crosses another 5,500 kilometers of the North Pacific seafloor. The active volcanoes of the Big Island are less than a million years old, but the Hawaiian Hot Spot has persisted for at least 75 million years. It has generated about 200 Hawaiian-type volcanoes, most of them now submerged, during its lifetime, and is still going strong.

The volcanic rocks, hot springs, and geysers of Yellowstone National Park in the western United States are related to another hot spot, whose track is the Snake River Plain in southern Idaho. The Azores, the Galapagos, and the Society Islands are examples of other volcanic islands formed by hot spots. Although the worldwide number of active hot-spot volcanoes is only about 50, some erupt frequently and, on average, account for about five eruptions on a yearly basis.

As new discoveries are made it becomes more and more evident that studies both of volcanoes and of plate tectonics are closely interwoven. The plate tectonic concept explains the location of volcanic belts, while the tracks of extinct hot-spot volcanoes reveal the direction and speed of ancient plate-motions. At the seams of the Earth, geology is not a dusty prehistoric science – it is alive with the challenge of understanding volcanic eruptions and great earthquakes, and of learning how their risks may be reduced.

Most hot-spot volcanoes are effusive; that is, they erupt streams of molten lava. In contrast, most volcanoes related to converging plates are explosive and erupt material that is largely hot but solid fragments. Chapter 3 explains more about the types of volcanic eruptions and why they differ so much from one another.

Table 2.1. *Number of volcanoes in the world considered to have been active in the past ten thousand years, listed by geographic area and country*

	No.
Africa	110
East Africa	98
Djibouti	3
Ethiopia	57
Kenya	17
Tanzania	6
Uganda	7
Zaire	8
North Africa	8
Chad	2
Libya	1
Sudan	5
West Africa	4
Cameroon	1
Equatorial Guinea	3
Antarctica	26
Asia (inland, north to south)	38
USSR	12
Mongolia	5
China	14
Korea	3
Burma	3
Vietnam	1
Australia	2
Europe (west to east)	21
France	1
West Germany	1
Italy	13
Greece	6
Middle East (north to south)	47
USSR	5
Turkey	11
Iran	3
Syria	6
Saudi Arabia	8
Yemen	12
South Yemen	2
North America (north to south to Panama Canal)	212
Alaska (except Pacific Rim) (USA)	12
Canada	20
USA (first 48 states)	69
Mexico	31
Guatemala	25
Honduras	2

(*continued*)

Table 2.1 (*cont.*)

	No.
North America, *continued*	
El Salvador	19
Nicaragua	22
Costa Rica	11
Panama	1
South America (north to south)	127
Colombia	13
Ecuador	10
Peru	10
Bolivia	15
Chile	75
Argentina	4
Atlantic Ocean (north to south)	105
Jan Mayen (Norway)	1
Iceland	62
Azores (Portugal)	8
Canary Islands (Spain)	7
West Indies	15
Netherlands	2
United Kingdom	3
France	2
Independent	8
Cape Verde Islands	1
Ascension Island (UK)	1
Tristan da Cunha Island (UK)	1
Bovet Island (Norway)	1
South Sandwich Islands (UK)	8
Indian Ocean (north to south)	144
Andaman Islands (India)	1
Indonesia	127
Comoros Islands	2
Madagascar	5
Reunion Island (France)	1
Amsterdam and St. Paul Islands (France)	2
Crozet Islands (France)	2
Prince Edward Islands (South Africa)	2
Kerguelen Islands (France)	1
Heard Island (Australia)	1
Pacific Ocean	450
Western and Northern Rim (south to north)	416
New Zealand	29
Kermadec Islands (New Zealand)	2
Tonga	4
Fiji	1
Matthew and Hunter Islands (France)	2
Vanuatu	10
Solomon Islands	7
Papua New Guinea	43
Philippines	51

Table 2.1 (*cont.*)

	No.
Northern Mariana Islands (USA)	7
Taiwan	3
Japan	77
Kurile Islands (USSR)	47
Kamchatka (USSR)	65
Aleutian Islands and Alaska Peninsula (USA)	68
Interior (north to south)	34
Hawaii (USA)	8
Revillagigedo Islands (Mexico)	2
Galapagos (Ecuador)	16
Western Samoa	3
Samoa (USA)	1
Mehetia (France)	1
Easter Island (Chile)	1
San Felix Island (Chile)	1
Robinson Crusoe Island (Chile)	1

After Simkin et al., *Volcanoes of the World*, 1981, and Supplement, 1982.

3

Volcanic eruptions

Mont Pelée, Martinique: May 8, 1902

The warning signs were all there; small but persistent earthquakes had shaken the city of Saint Pierre for several weeks, and the moderate explosive eruptions at the summit of Mont Pelée seemed to be increasing, keeping the citizens constantly aware of the volcano that towered above them (Fig. 3.1). Then suddenly the mountain unleashed an enormous glowing avalanche that swept down and demolished the port of Saint Pierre in the greatest volcanic catastrophe of the twentieth century (Fig. 3.2).

The tragedy was compounded by the fact that because elections were about to be held on the island, the governor had encouraged people not to flee from Saint Pierre. He felt the opposition party was favored by the people in the countryside and did not want any votes to slip away. In a few awful moments, voters and governor alike were incinerated by the ash flow.

Purser Thompson, who was on the steamship *Roraima* in the harbor, lived to tell this vivid tale:

> I saw Saint Pierre destroyed. It was blotted out by one great flash of fire. Nearly 40,000 people were all killed at once. Of eighteen vessels lying in the roads only one, the British steamship Roddam, escaped, and she, I hear, lost more than half on board. It was a dying crew that took her out. Our boat, the Roraima, of the Quebec Line, arrived at Saint Pierre early Thursday morning. For hours before we entered the roadstead we could see flames and smoke rising from Mont Pelée. No one on board had any idea of danger. Capt. G. T. Muggah was on the bridge and all hands got on deck to see the show. The spectacle was magnificent. As we approached Saint Pierre we could distinguish

Figure 3.1. Saint Pierre and Mont Pelée on the Caribbean island of Martinique before the 1902 eruption. (From F. Royce, The Burning of St. Pierre, [Chicago: 1902], Continental Publishing Company)

the rolling and leaping of the red flames that belched from the mountain in huge volumes and gushed high in the sky. Enormous clouds of black smoke hung over the volcano. When we anchored at Saint Pierre I noticed the cable steamship Grappler, the Roddam, three or four American schooners and a number of Italian and Norwegian barks. The flames were then spurting straight up in the air, now and then waving to one side or the other for the moment, and again leaping suddenly higher up. There was a constant muffled roar. It was like the biggest oil refinery in the world burning up on the mountain top. There was a tremendous explosion about 7:45 o'clock, soon after we got in. The mountain was blown to pieces. There was no warning. The side of the volcano was ripped out, and there hurled straight toward us a solid wall of flame. It sounded like thousands of cannon. The wave of fire was on us and over us like a lightning flash. It was like a hurricane of fire. I saw it strike the cable steamship Grappler broadside on and capsize her. From end to end she burst into flames and then she sank. The fire rolled in mass straight down upon Saint Pierre and the shipping. The town vanished before our eyes and then the air grew stifling hot and we were in the thick of it. Wherever the mass of fire struck

Figure 3.2. Saint Pierre was totally destroyed by the eruption of Mont Pelée. Twenty-nine thousand people were killed; only two survived. (Photograph by Underwood and Underwood, from the Library of Congress)

the sea the water boiled and sent up vast clouds of steam. The sea was torn into huge whirlpools that careened toward the open sea. One of these horrible hot whirlpools swung under the Roraima and pulled her down on her beam ends with the suction. She careened way over to port, and then the fire-hurricane from the volcano smashed her and over she went to the opposite side. The fire wave swept off the masts and smokestack as if they were cut with a knife. Captain Muggah was the only one on deck not killed outright. . . . The blast of fire from the volcano lasted only a few minutes. It shriveled and set fire to everything it touched. Burning rum ran in streams down every street and out onto the sea. This blazing rum set fire to the Roraima several times. Before the volcano burst the landings of Saint Pierre were crowded with people. After the explosion not one living being was seen on land. Only twenty-five of those on the Roraima, out of sixty-eight, were left after the first flash.*

Official figures indicate that about 29,000 people were killed. Many refugees from the less destructive eruptions, which preceded by about a week the main blast of May 8, had sought haven in the city even as others were making plans to leave. The destruction was so complete that there was no one left to count the dead. Only two survivors in town lived through the fiery gas cloud that swept down on Saint Pierre; one was a prisoner in a dungeon-like jail and the other was at the outer edge of the devastated area.

Contrary to Thompson's first impression, Mont Pelée was not blown to pieces. The old crater rim was partly destroyed and the former crater area was not recognizable, but the flanks of the mountain were unchanged except for the scour and fill of the hot avalanches. Purser Thompson and his few fellow survivors did not linger to see the aftermath of the great eruption, and during such a catastrophe it would be quite easy to believe that the entire mountain was blown to destruction while concealed within such a violent glowing cloud of hot gas and rock debris.

VOLCANIC ERUPTIONS

Why do volcanoes erupt? Why are some eruptions explosive, while others are quieter emissions of rivers of molten lava? Why does an eruption stop, and why is the time between eruptions so variable from one volcano to another? It is easier to describe the character of various types of volcanic eruptions than to answer these questions, but some possible solutions are emerging from the studies of active volcanoes by scientists in many countries.

Volcanoes erupt for two basic reasons: Magma deep in the Earth is generally less dense than the solid rocks surrounding and overlying it, and it tends to rise toward the Earth's surface from the buoyant force

---

* "The Destruction of the Roraima," *Frank Leslie's Popular Monthly,* July 1902.

Figure 3.3. If a cork and a ball bearing are submerged in a jar of honey, the cork will slowly rise and the ball bearing will sink. Both the rising and sinking are caused by gravity; in a spaceship under zero gravity neither object would move. Batches of magma rise in the Earth's gravity field because they are less dense than the surrounding rocks.

of gravity (Fig. 3.3). As magma nears the surface, its dissolved gases boil out of the molten rock (Fig. 3.4). The force of this expansion propels the molten lava, or hot but solid lava fragments, from the vent.

EXPLOSIVE ERUPTIONS

If the gas content is high and the magma thick and viscous, sudden release of confining pressure allows the gases to boil explosively from the magma. This sudden bursting tears the magma into hot fragments that jet upward or blast outward from the vent. The direction of travel of the mixture of hot ash (fine volcanic rock particles), larger rock fragments, and expanding gases depends in part on the initial shape and direction of the vent, and even more critically on the density of the mixture of hot rock particles and gas.

If the ash cloud is lighter than air, it rises into a dark roiling cloud with many turbulent cells. If the mixture of hot rock particles and gas is denser than air, it moves downslope as a high-speed avalanche like the one that destroyed Saint Pierre. These glowing avalanches, also known as *nuées ardentes,* ash flows, and pyroclastic flows, are some of the most awesome and destructive of nature's arsenal. They are hot – sometimes as much as 700° C – and travel at speeds that may exceed 100 kilometers per hour. They annihilate all life and most obstacles in their path.

EFFUSIVE ERUPTIONS

If the gas content in the magma is low and the magma is of relatively low viscosity, the gases boil out less violently. The fiery lava fountains

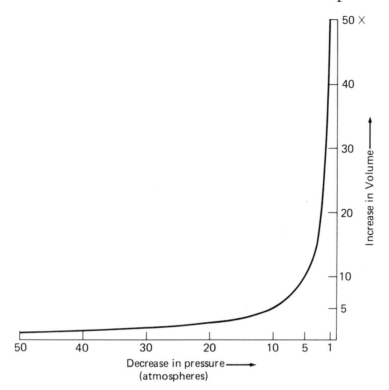

Figure 3.4. Graph showing the approximate volume of volcanic gases at a constant high temperature and varying pressure. For each 10 meters of depth below sea level, or about 4 meters of depth below ground level, pressure increases by 1 atmosphere. For example, volcanic gas bubbles in magma at a depth of 36 meters below ground surface (10 atmospheres) would expand approximately 10 times in volume as they approach the surface (1 atmosphere).

that characterize many Hawaiian volcanic eruptions are examples of this effusive type of eruption. Sometimes, if the gases have escaped before the lava reaches the surface, the lava will quietly well out of the vent as an incandescent stream of molten rock.

Naturally, there are gradations between extremely explosive and quietly effusive eruptions. Possibilities are further multiplied if surface- or groundwater is present. A few hundred meters below the Earth's surface, the pressure of overlying rock increases the boiling point of water to more than 200° C. If groundwater at these depths is heated to 200° C by nearby magma and the overlying pressure is suddenly reduced, as happened at Mount St. Helens, the hot water will flash to steam in a massive hydrothermal explosion. Any volcanic eruption that suddenly mixes groundwater and molten rock in about equal quantities will also create huge steam blasts.

Both these types of groundwater explosions can occur at volcanoes with viscous, gas-rich magma, thus increasing the overall violent effects.

They are also possible, though much more rare, at normally effusive volcanoes like Kilauea Volcano in Hawaii.

SUBMARINE ERUPTIONS

Submarine eruptions at great depth are usually effusive rather than explosive because the water pressure retards boiling or rapid gas expansion. Conversely, in shallow water, eruptions tend to be more violent because of the rapid generation of steam from the thorough mixing of magma and sea water at low pressures.

One dramatic exception to this general rule is seen when the smooth-surfaced lava flows in Hawaii – called "pahoehoe flows" – enter the sea. These flows usually have traveled long distances and in the process have lost most of their internal gases; they slip into the sea with remarkably little generation of boiling steam. The smooth surface of these flows, and the generation of a thin layer of steam between the flow surface and the seawater, retards the heat transfer from the molten lava to the water and prevents explosive boiling. The same effect is seen when a drop of water falls onto the surface of a hot skillet on a stove – instead of flashing into steam, the droplet skitters and dances around the skillet on a small cushion of steam.

VOLCANIC FIRE

Volcanic eruptions do not involve combustion on any significant scale. A small amount of hydrogen gas is dissolved in magma, but when it burns with atmospheric oxygen to form water only a tiny amount of energy is released compared to the heat content of the molten rock. Although a fume cloud or ash cloud from a volcano is sometimes referred to as "volcanic smoke," that is a misnomer; the fire in volcanic eruptions is incandescent rock, not the fire of combustion of fuel with oxygen. However, secondary fires often do occur when brush and trees are ignited by molten lava or hot rock fragments.

Even though it is common to compare the energy release in a large eruption to the energy yield of a nuclear bomb, there are no nuclear reactions involved in volcanic eruptions. The energy and power of a volcanic eruption is contained in the enormous store of calories in the 900° to 1,200° C temperature of magma. The sudden conversion of that heat energy by the explosive boiling of volcanic gases (carbon dioxide, steam, and sulfur dioxide) and adjacent groundwater or surface water generates the power. Conversely, an eruption stops when the rapid release of energy from the expanding gases exceeds the rate of energy replacement from new magma rising from depth. It is similar to blowing up a balloon until it bursts; the slow storage of energy is suddenly released when the balloon pops.

REPOSE TIMES

Sometimes the slow accumulation of magma, and the sudden release of volcanic energy in relatively short periods of eruption, is remarkably rhythmic. The spectacular lava fountains on the flank of Kilauea Volcano in Hawaii during the past few years are good examples of cyclical eruptions. From January 1983 to July 1986 there were 47 episodes of high, spectacular lava fountaining from the same general vent area. These eruptive episodes, each lasting for several hours, produced about 10 to 15 million cubic meters of lava, and were separated by repose times of about a month between episodes.

Such regularly recurring eruptions are the exception, however. Repose periods differ greatly among various volcanoes, and an individual volcano can have repose periods of different times as well. Stromboli has an average repose time of about 15 to 20 minutes between its small eruptions, while other volcanoes, such as El Chichón in Mexico, are quiet for more than 1,000 years between eruptions. Asama Volcano in Japan has erupted thousands of times since its first recorded eruption in A.D. 685, and since 1900 its shortest repose times have been less than one day, and the longest, five years.

Repose times vary among volcanoes because of the enormous diversity of volcanic types, rock types, surface environments, and stages of volcanic growth. An individual volcano's repose periods may differ because the rate of magma movement from depth is variable, and because the strength of the volcanic edifice changes with each eruption. For example, an eruption that causes a fracture zone in the flank of a volcano may weaken the structure, and this may be followed by more frequent but smaller eruptions. Imagine the balloon analogy again; if the balloon is inflated at different rates or the strength of the balloon's skin is uneven, the time it takes to reach the bursting point will also change.

Because an eruption ceases when energy is released faster than it is replaced by new magma from depth, extremely violent eruptions are generally short-lived, whereas steady, low-volume lava eruptions, or a series of small explosive eruptions, may continue for many years – in the case of Stromboli, for centuries.

EXPLOSIVE VERSUS EFFUSIVE ERUPTIONS

Some volcanoes are predominantly explosive in their habits while others are more effusive. This difference involves at least four important factors: the viscosity of the magma, the amount of gases dissolved in the magma, the suddenness with which the pressure on the magma is reduced as it nears the Earth's surface, and the amount of heated groundwater beneath the volcano.

The viscosity of magma ranges from a liquid of about the consistency of honey to nearly solid material. High temperatures and low silica content, as generally occur in basaltic magma, lead to low viscosities; lower magma temperatures and higher silica content, as generally occur in more silica-rich dacitic magma, lead to high viscosities (Part II discusses the composition of magma and volcanic rocks). Below temperatures of about 600° to 700° C, volcanic rock is still hot, but essentially solid.

The ease with which gases can boil out is largely controlled by the viscosity of the magma. At low viscosities the gases escape relatively rapidly and form lava fountains or boiling lava ponds; at high viscosities the pressure of gases forming bubbles in the molten rock build up to the point at which they shatter the magma into hot fragments. Gunpowder is similar in this respect (except that it involves combustion); it burns if unconfined, but explodes if wrapped in paper firecrackers or encased within a bullet cartridge.

Because the boiling gases dissolved in magma propel volcanic eruptions, explosiveness increases with gas content. Basalt magmas have about 0.5 to 1.0 percent (by weight) of dissolved volcanic gases, while the more silica-rich magmas generally have increasing amounts of dissolved gases, up to as much as 5 to 6 percent.

The source of these additional volcanic gases in magmas more common to subduction-related volcanoes is not clearly understood. It may be related to the incorporation of seafloor sediments from the subducting plate into the magma formed at depth. Seafloor sediments have high water (and often high carbon dioxide) contents. Alternatively, the process of forming more silica-rich magmas may add or concentrate the dissolved gases. Whatever the process, the more gas dissolved in the magma, the more potential there is for large explosive eruptions.

Finally, the rate at which the pressure on the magma is lowered as it ascends toward the surface also has an important bearing on the explosiveness of a volcanic eruption. Slow ascent allows the gases – especially carbon dioxide, which comes out of solution with magma at greater depth than water or sulfur gases – to separate from the magma and leak away harmlessly to the surface. Rapid uprising, or very sudden depressurization, as in the case of the avalanche at Mount St. Helens, will trap the gases in the magma until the final moment when they burst forth like an exploding steam boiler.

Volcanic eruptions are extremely variable in size and character, but they have two fundamental common principles: Gravity forces the lighter magma from within the Earth to the surface, and gentle to explosive boiling of gases determines the type of eruption.

The life-span of a volcano is also highly variable – from a few years to millions of years – but can be divided into stages. Those life stages are the subject of the next chapter.

4

Life stages of volcanoes

Michoacán, Mexico: February 20, 1943

Michoacán province shook with a violent earthquake. At a farm near the village of Parícutin the ground swelled as much as two meters as, with a roar of thunder, a long fissure tore Dionisio Pulido's cornfield apart. From a hole at one end of the fissure smoke, steam, and bursts of red-hot cinders poured into the air, signaling the birth of a new volcano.

The Michoacán area of south-central Mexico is dotted with hundreds of old cinder cones built by similar eruptions, but until 1943 only one, Jorullo, which grew from 1759 to 1775, had erupted in historic time. The farmer who was plowing his field near the village of Parícutin had noticed that the ground underfoot felt warm and occasionally a wisp of smoke curled into the air from a small depression, but nothing in his experience led him to suspect that something dramatic was about to happen.

Over the first few hours, the earthquakes continued and hot ash emissions increased in volume, including red-hot stones thrown 10 meters in the air. That night the new volcano was a dazzling sight, with incandescent bombs hurled as high as 500 meters and lightning flashing through the ash column (Fig. 4.1). As the rocks and cinders fell back to earth they began to build up a steep cone; by the next morning it was 10 meters high, and within a few hours it had grown to nearly 40 meters from the voluminous eruption.

The new volcano was given the name Parícutin after the nearby town; ironically, lava flows from the volcano were to destroy its namesake town just a few months later. During the early months Parícutin grew remarkably fast. After the first week it was 140 meters high; after ten weeks, 270 meters. Continuous explosions – 30 to 40 a minute – emitted steam and dust clouds rising to 1,500 meters and rock fragments thrown

Figure 4.1. Lightning bolts are common in and around turbulent ash clouds from volcanic eruptions. This five-minute time exposure of Cerro Negro Volcano in Nicaragua was taken in 1971. The eruptions of Parícutin were similar to those at Cerro Negro. (Photograph by Ingeniero Franco Penalba)

as high as 500 to 1,000 meters. The eruption continued in this manner for months, with the cone growing both in height and width. The surrounding area was devastated, not only by the encroaching pile of the volcano but also by the fall of ashes that smothered crops and crushed houses for miles downwind.

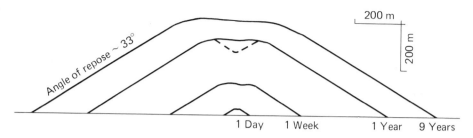

Figure 4.2. Profiles showing the growth of Parícutin Volcano in Mexico from 1943 to 1952.

During this period a thick lava flow would occasionally issue from a boca, an opening near the base of the cone, and spread sluggishly over the surrounding land. Then, in June, a new phase of the eruption began. After a long night of even more violent explosions and earthquakes, a crack opened at one edge of the crater in the top of the cinder cone, and liquid lava poured out. It flowed down the side of the new mountain and during the next months inundated the now abandoned countryside. This was the only time in the course of Parícutin's life that lava flowed from the crater; the voluminous flows to come all issued from bocas.

Later in the year a new cone named Sapichu opened on Parícutin's side and became violently active, with liquid lava flows and explosions of ash and bombs. The main crater was less active now, erupting mostly steam and gas. As the secondary cone died, a series of bocas opened on the other side of the mountain with the most spectacular lava flows of the eruption. Massive flows poured around the base of the cone and engulfed two towns that had been untouched until that time, including the village of Parícutin.

The eruption continued but with gradually lessening fury for the next nine years. The cone grew to most of its full height of 410 meters in the first year (Fig. 4.2). After that, the instability of the loose pile of material caused slides and slumping so that the volcano grew more in volume than in height. Lava still flowed from a series of bocas at different locations around the cone, but generally stayed on top of older flows and did not appreciably enlarge the outlines of the lava field (Fig. 4.3).

In March 1952 the eruption ceased almost as suddenly as it began; Parícutin is now a silent black cone on the plain of Mexico. In the region near Parícutin there are many prehistoric cinder cones that look as if they have erupted just once. It seems probable that the entire active life of Parícutin Volcano may have occurred in the few brief years from 1943 to 1952.

Figure 4.3. The cathedral tower rises like a gravestone above the buried town of San Juan Parangaricutiro. The thick block lava flow from Parícutin advanced slowly enough so that no one in the town was killed. (Photograph by Katia Krafft)

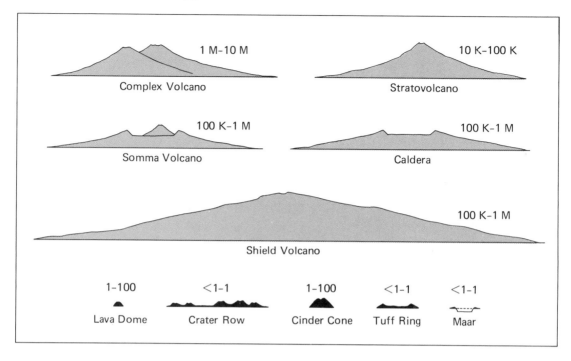

Figure 4.4. Relative sizes, shapes, and life-spans of various types of volcanoes. The shaded profiles are exaggerated vertically by a factor of 2; the black profiles, by a factor of 4. The numbers are the approximate span of years during which this type of volcano is generally active. M = 1 million and K = 1,000. (Size and shape profiles after Simkin et al., Volcanoes of the World, *p. 10)*

VOLCANIC LIFESPANS

For volcanologists, watching the birth and death of Parícutin was like seeing a speedup of geologic time. Being able to observe the changing life stages of that volcano compressed into nine years gave valuable clues for the study of other volcanoes whose lives may be a thousand – or a million – times as long (Fig. 4.4). The average life-span of a Hawaiian

Figure 4.5. Schematic diagrams of the life stages of Hawaiian volcanoes. Loihi seamount is an example between stages 1 and 2; Surtsey Volcano in Iceland, stage 3; Mauna Loa, stage 4; and Mauna Kea, stage 5. The Island of Oahu has been through all or part of stage 7; Midway Island is an example of stage 8; and several seamounts in the Emperor chain, northwest of Hawaii, stage 9. Vertical scale has been exaggerated in these cross sections to show detail. The alkalic lavas in stages 1, 5, and 8 are largely basalts high in sodium and potassium. (Modified from D. W. Peterson and R. B. Moore, Volcanism in Hawaii *[Professional Paper 1350], [U.S. Geological Survey, 1987], p. 169)*

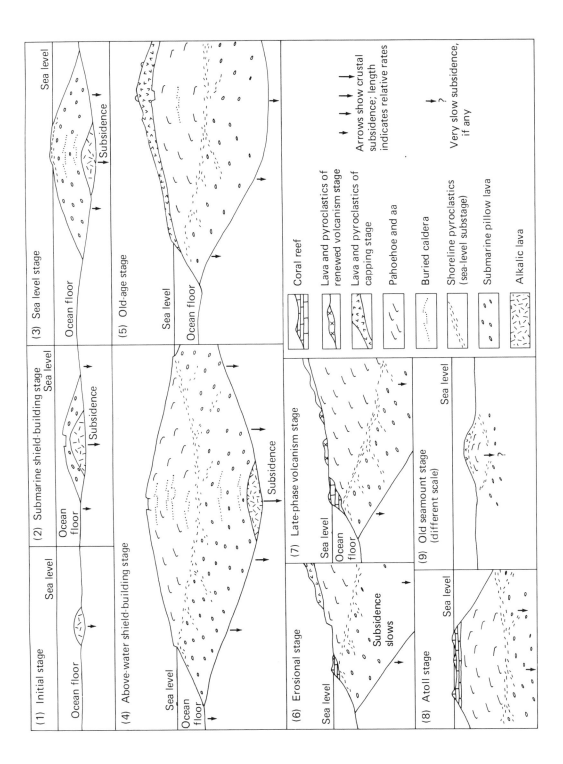

(1) Initial stage

Sea level

Ocean floor

(2) Submarine shield-building stage

Sea level

Ocean floor

Subsidence

(3) Sea level stage

Sea level

Ocean floor

Subsidence

(4) Above-water shield-building stage

Sea level

Ocean floor

Subsidence

(5) Old-age stage

Sea level

Ocean floor

Subsidence

(6) Erosional stage

Sea level

Subsidence slows

(7) Late-phase volcanism stage

Sea level

Ocean floor

(8) Atoll stage

Sea level

(9) Old seamount stage (different scale)

Sea level

?

Coral reef

Lava and pyroclastics of renewed volcanism stage

Lava and pyroclastics of capping stage

Pahoehoe and aa

Buried caldera

Shoreline pyroclastics (sea-level substage)

Submarine pillow lava

Alkalic lava

Arrows show crustal subsidence; length indicates relative rates

Very slow subsidence, if any

?

43

volcano is about 500,000 to a million years, and some volcanic centers stay intermittently active for more than 10 million years.

Because Hawaiian volcanoes have been thoroughly studied, and because individual Hawaiian volcanoes are in different stages of growth and decline, their typical life stages have been fairly well defined and make good examples for use in this discussion.

LIFE STAGES OF A HAWAIIAN VOLCANO

Hawaiian volcanoes are born on the deep ocean floor, below about 5 kilometers of water, or perhaps on the submarine flank of an earlier volcano (Fig. 4.5). As basaltic lava is erupted deep beneath the surface of the sea, the pressure of the overlying water prevents boiling of much of the gases in the magma and of the water in contact with the hot erupting lava. Even though common sense suggests that seawater would quench a submarine volcanic eruption, or that the struggle between cold water and hot lava would be violent, the evidence from deep submarine volcanic rocks is that they pour out of their vents much as in the most quiet Hawaiian type eruptions above sea level, and that they flow considerable distances before cooling and solidifying.

The ability of lava to stay molten even when it is submerged in water results from the excellent insulating property of the skin of solidified volcanic rock that forms on the surface of the molten lava. As lava streams flow from submarine vents, a rapidly solidified skin forms sacklike covers to contain the fluid rock. These sacks inflate and then rupture; lava flows from the tear and is quickly covered with new solid skin to form new sacks in an advancing flow of submarine lava. The accumulated layers of these solidified sacks has been described often as looking like a pile of pillows, and the name has stuck. Pillow lava formed on the deep sea floor is one of the more common rock types on Earth; most of it, however, is hidden beneath the sea and is only observed where land has been uplifted.

As layers of pillow lava grow upward and approach the surface of the sea, a Hawaiian volcano enters a new and violent life stage. In shallow water, the gases in erupting magma and the water heated by contact with the molten lava expand rapidly. Explosive boiling rips the lava into fragments, which it hurls into jets above the churning sea. The fragments fall back to form a ring of debris around the boiling vent. Diamond Head on Oahu and Molokini, a small islet south of Maui, were formed this way, but at present no Hawaiian volcano is in this life stage.

SURTSEY VOLCANO

The eruption that formed Surtsey Island in Iceland in 1963–7 was probably a close analogy to the transition of a Hawaiian volcano from a submarine to an island volcano. The Surtsey eruption was first noticed at dawn from a fishing boat as dark smoke on the horizon. Rushing to aid what they thought to be a ship on fire, the fishermen soon recognized the jets of volcanic ash and steam rising from the sea. By afternoon nearly continuous explosions issued from a 500-meter line across the sea surface, and dark ash plumes fell back into the water as a huge white steam cloud boiled up for thousands of meters.

The sea bottom near Surtsey is 130 meters deep, and presumably the eruption started quietly days to weeks before it built up to near the sea surface where the explosions began. Two days before the eruption broke the surface, a marine research ship measured unusually warm temperatures in the seawater in the area, and people in a coastal village had noticed the rotten-egg smell of hydrogen sulfide.

During the first week of explosive activity, the low island of volcanic debris grew nearly continuously, with ash and steam rising in a turbulent column to heights of 8 to 10 kilometers. Ten days into the eruption Surtsey was an island 900 meters long and 650 meters wide, with an arcuate crater rim about 100 meters high.

The steam explosions varied from single events separated by a few minutes, to continuously uprushing columns of steam and ash lasting hours. Glowing red fragments of pasty lava called "volcanic bombs" and hundreds of lightning bolts lit up the eruption column at night. Surtsey, from Surtur, legendary Nordic giant of fire, was aptly named.

As the island of loose volcanic fragments grew higher, seawater could no longer reach the crater. By 1964, about six months after Surtsey appeared, lava fountains and flows became the dominant type of eruption. These formed a hard cap of solid rock over the loose volcanic material and armored the island from the relentless attacks of the stormy North Atlantic.

By 1967 the birth of Surtsey was over. The total volume of volcanic ash and lava was about 1 cubic kilometer; less than 10 percent of that volume rose above sea level. The eruption had three phases: the quiet underwater buildup from 130 meters to just below sea level; the violent initial island-building eruptions from explosive boiling of gases, mostly steam from contact seawater; and after seawater was blocked from the vent, the quiet effusion of lava flows.

Surtsey may erode to a shallow shoal before it erupts again; many volcanoes in Iceland have only one eruption or long repose periods of hundreds to thousands of years between eruptions. In that way Surtsey differs from Hawaiian volcanoes; nevertheless, its birth from the sea was probably similar to the initial island-building eruptions of volcanoes in other parts of the world. Good examples of the next three life stages of volcanoes can all be found on the Big Island of Hawaii: Kilauea in the vigorous, youthful stage; Mauna Loa in the mature, shield-building stage; and Mauna Kea declining into old age.

HAWAIIAN VOLCANOES

Building tongue on tongue of lava above the sea, Kilauea Volcano has grown to 1,200 meters above sea level in about 100,000 years. Presently the most vigorous of Hawaiian volcanoes, Kilauea was in almost continuous eruption with an active lava lake in its summit caldera during the hundred years before 1924. Following the draining of the lava lake in that year, there were, through 1990, 41 short- to long-lived eruptions from the summit and flanks of Kilauea. This very active period in the life of a Hawaiian volcano is called its "vigorous shield-building stage."

Mauna Loa Volcano on the island of Hawaii is Kilauea's giant neighbor. Its shield gently rises to more than 4,000 meters above sea level. With its base more than 5,000 meters below sea level and constructed of 40,000 cubic kilometers of basalt, Mauna Loa is the world's largest volcano. Its output of lava in the past 150 years has been about the same as Kilauea's production, but individual eruptions are less frequent and more voluminous. Mauna Loa probably emerged from the sea 500,000 years ago and began life on the seafloor another 500,000 years before that. It is still vigorous, but it is possibly approaching the end of its shield-building stage.

Mauna Kea Volcano has not erupted for more than 3,000 years but is only 36 meters higher than Mauna Loa, its younger neighbor on the Hawaiian volcanic chain. Not only is Mauna Kea older and a bit higher than Mauna Loa, its eruptive habits have changed significantly. Instead of producing numerous lava flows from its summit or along weak zones on its flanks, it has had many fewer eruptions, most of which have built large cinder cones. The cones are concentrated at the summit but are also found scattered on the flanks of Mauna Kea. These cinder cone eruptions were probably much like Parícutin's: forming cones rapidly in one or more eruptions, with large separations in time and space between them.

Following a long period of old age, Hawaiian volcanoes begin to erode away and sink slowly beneath the sea. Actually, all volcanoes on the island of Hawaii are slowly sinking from their own great weight on the Earth's crust, but Kilauea and Mauna Loa are still growing upward more rapidly than they are sinking.

Streams cut deep valleys into the older volcanoes, especially on their wet windward sides, and the ocean cuts away the islands' waveswept shores into steep cliffs. If sea level is steady, large reefs form over periods of thousands of years, especially on the lee side of the islands. On the older islands like Oahu, the mountain ridges are just the harder inner skeletons of the ancient volcanoes, dormant now for 2 million years or more.

LATE-PHASE ACTIVITY

The story of the life stages of Hawaiian volcanoes would now be nearly over except for a last fling of activity, called the posterosional stage. This late-phase activity is not well understood. It is small in volume and long in repose between eruptions, and the composition of the lava is different from the main growth stage of Hawaiian volcanoes.

Some volcanoes of this late phase grew in what is now downtown Honolulu. Most of the volcanic rocks in the mountains behind Honolulu were formed about 2 to 3 million years ago in the main stage of the Hawaiian volcano that formed the eastern part of Oahu. However, the young volcanic features in Honolulu, such as Diamond Head and Punchbowl, were formed by posterosional volcanic activity that occurred about 100,000 years ago. It is still possible, though unlikely, that renewed volcanic eruptions will occur again on Oahu.

After the posterosional volcanic activity Hawaiian volcanoes continue to erode to sea level and slowly subside beneath the seas as they drift northwestward on the Pacific plate. Fringing coral reefs keep growing upward and form atolls. Where there had once been some great volcano, there is now only a shallow lagoon. Drilling at Midway Atoll has proven that volcanic rocks lie beneath the coral capping.

LOIHI

Loihi seamount is the youngest Hawaiian volcano, still hidden by the Pacific Ocean. Its summit is 1 kilometer below sea level and 30 kilometers southeast of the island of Hawaii. Earthquake swarms centered beneath Loihi in 1971–2 and in 1975, and fresh lava samples dredged by oceanographic ships from its summit and sides indicate that Loihi is very much alive. One surprise finding from detailed echo soundings of the submarine topography around Loihi (which means "the long one") is that collapse craters and rift zones apparently exist on this young submarine volcano.

No one is sure how old Loihi is. Its age can be estimated only by its present size and the known growth rate of Kilauea Volcano, on whose south flank Loihi is found. Kilauea has been growing at a rate of about one-tenth of a cubic kilometer per year for the past few decades, and the volume of Loihi is about 100 cubic kilometers. This implies that Loihi is only about 1,000 years old, but perhaps young Hawaiian volcanoes grow slowly compared to their later growth rate. It is clear only that there is much yet to be learned about the youthful stages of Hawaiian volcanoes.

OTHER VOLCANOES

The life stages of subduction volcanoes are more complex than those of Hawaiian volcanoes. Their activity may continue for millions of years, but their explosive habits sometimes destroy as well as create their giant cones. One possible history might begin with a cinder cone, eventually growing into a high stratovolcano whose summit finally collapses into a giant caldera. Some calderas appear to rebuild into a more complex stratovolcano, only to collapse again. As long as the subduction process continues in the same general location, subduction volcanoes may continue to erupt, forming many stages and generations during their long and complex lifetimes.

Single short-lived volcanoes like Parícutin form many small cones scattered over the landscape, whereas longer-lived explosive volcanoes form isolated large stratovolcanoes. If magma rises often enough, it probably reuses its old conduits and forms a long-lived volcano. However, if the period between pulses of magma is long enough for the conduit from depth to solidify, a new conduit may form a separate, short-lived volcano.

The processes of erosion finally subdue even the longest-lived volcanoes. The elements of which they were composed are recycled again and again in the restless motion of the Earth's tectonic plates.

5

Volcanoes in the solar system

Pasadena, California: March 4, 1979

Excitement was high among the planetary scientists who gathered at monitors to see the first historic pictures of Jupiter and its moons as they were beamed back to Earth by Voyager I. At the Jet Propulsion Laboratory in Pasadena, Linda Morabito was processing an image of Io, one of the moons of Jupiter, when she noticed something unusual.

Her research involved determining the exact orbit of Io and the path of the Voyager I spacecraft. To do this she had to intensify the image of Io to reveal two faint stars behind the Jovian moon; the location of Io with respect to those stars would then provide an exact fix of the relative position of Io and Voyager I. Not only did the dim stars show up on the overexposed image, but also a strange, thin crescent appeared above the bright eastern limb of Io.

Morabito concluded that this unexpected feature, reaching 280 kilometers above the surface, might be a plume of gas or particles from an active volcano on Io. Other images from Voyagers I and II proved this interpretation to be correct, and active volcanism suddenly became a planetary process – not just a phenomenon unique to Earth (Fig. 5.1).

The discovery was not a complete surprise, however. A few months before Voyager I reached Jupiter, S. J. Peale, P. Cassen, and R. T. Reynolds, scientists with the University of California and NASA, predicted that Io might have a molten interior. By studying the orbit of Io around Jupiter, they concluded that tidal strains on Io caused by the enormous mass of Jupiter could generate enough friction to melt Io's interior and keep it that way. They stated that "one might speculate that widespread and recurrent surface volcanism would occur...." The prediction of active volcanism on Io based on theoretical calculations, and

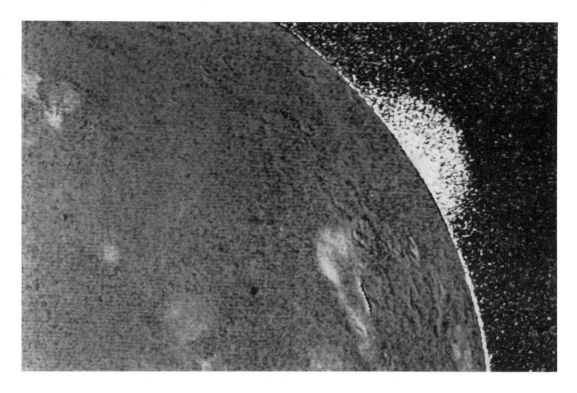

Figure 5.1. A cloud of particles and gas 200 km high is erupted above the orange-colored surface of Io, a moon of Jupiter. Several active volcanoes, thought to be composed of sulfur and sulfur compounds, have been observed on Io. The eruption clouds reach such great heights because of Io's low gravity and thin atmosphere. (Voyager 1 image, NASA, JPL)

its subsequent confirmation by observation, is an example of scientific research at its best.

PLANETARY GEOLOGY

Ask a group of geologists what the most pervasive geologic process on Earth is; many will answer erosion, and some will say plate tectonics. Planetary geologists, if asked the same question about the solar system, would be more likely to answer volcanism. Erosion by streams and ocean waves can take place only in worlds with abundant water, and Earth appears to be the only water planet in our solar system. In contrast, volcanoes or volcanic rocks, some in strange forms, appear to be important features on most of the rocky planets and moons that have been explored to date.

Figure 5.2. The side of the Moon seen from Earth. The densely cratered highlands (lighter areas) were mainly formed before 3.9 billion years ago. Flooding of the mare ("seas") with flows of basaltic lava occurred after 3.9 billion years ago when the rate of meteorite bombardment rapidly decreased. (Photograph by the Pic du Midi Observatory, France, courtesy of NASA)

THE MOON

Since telescopes first revealed the numerous craters on the moon, scientists have argued about whether they are of volcanic origin or are caused by impacts of meteors (Fig. 5.2). Space probes and landings on the Moon have resolved that argument in favor of impact scars, but it is evident that lunar volcanism has been important in the distant past. The "seas" on the Moon are great basins, called "Maria," filled with basaltic lava flows; most of these erupted between 3 billion and 4 billion years ago – longer ago than most rocks on Earth. These lunar lava flows must have been very fluid, and were erupted in volumes that greatly

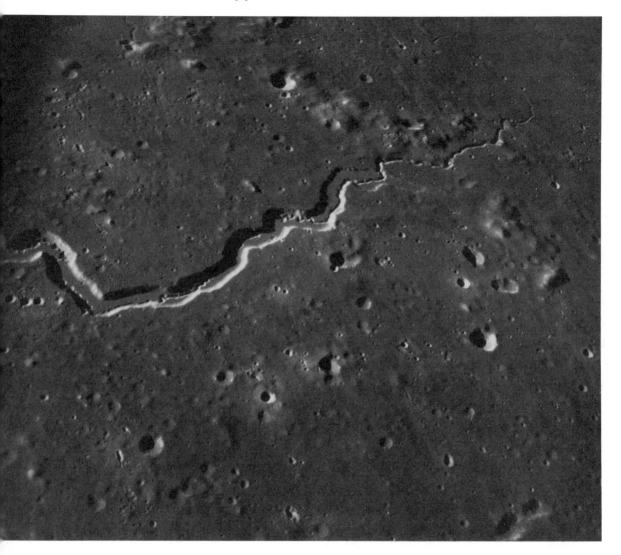

Figure 5.3. A sinuous rille, Schroters Valley, winds across the surface of the Moon. This large valley, up to 9 km wide, extends across the Aristarchus Plateau and out into the Oceanus Procellarum, one of the "seas" of the Moon. Analysis of samples collected by astronauts, and of photographs from space, has established that the lunar seas are great basins filled with basaltic lava, and that many of the rilles are huge lava-flow channels. (NASA photograph)

exceed the rates of even the largest Hawaiian or Icelandic eruptions. Some lunar flows are tens of kilometers wide and hundreds of kilometers long. Mare Imbrium, the largest "sea" on the moon, is covered by basaltic flows more than four times larger in area than the Columbia lava plateau in the northwestern United States. Because these ancient lavas are undisturbed by any significant erosion or mountain building, their general features, such as flow channels, are still well preserved (Fig. 5.3).

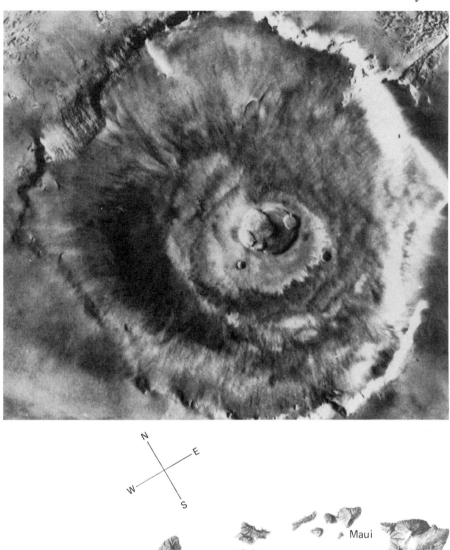

Figure 5.4. Olympus Mons, a giant volcano on Mars, is 25 km high and 600 km wide. The above-water portion of Mauna Loa, Earth's largest volcano, would nearly fit into the 80-km-wide caldera of Olympus Mons. The state of Hawaii is shown at the same scale for comparison. (Mariner 9 image mosaic, NASA, JPL; state of Hawaii map from U.S. Geological Survey)

MARS

Mars, too, has extensive lava fields of what appear to be basaltic flows, and also some huge shield volcanoes. Olympus Mons rises 25 kilometers from its base and is 600 kilometers in diameter (Fig. 5.4). Los Angeles and its suburbs would fit within the huge 80-kilometer-wide caldera of

Olympus Mons. The outer slopes of this giant shield volcano slope gently down to a great 2- to 5-kilometer-high cliff that surrounds the base of Olympus Mons. Landslides may have helped form this cliff, but its origin is still a matter of lively debate.

Perhaps Olympus Mons is the Martian equivalent of the Hawaiian Hot Spot, but with no plate tectonics operating on Mars to stretch the volcanoes into a chain, the single volcano probably kept growing in the same spot to its giant proportions. Based on the number of meteor crater scars that pockmark its slopes, the last volcanic activity on Olympus Mons occurred about 200 million years ago.

Space-probe photos of Mars show other volcanic features, including enormously wide volcanoes of even lower relief than shield volcanoes on Earth. One of these is 1,500 kilometers across and less than 10 kilometers high. Flows of what appears to have once been very fluid lava meander down its gently sloping flanks. Some large Martian volcanoes seem to be surrounded by ash deposits that have been partly eroded away, and many smaller volcanoes of unknown type are scattered across the red planet. Mars may once have had running water, but its present features show a barren land of ancient volcanoes swept by violent sandstorms.

VENUS

There is evidence for both ancient and active volcanoes on Venus, but that evidence is largely indirect. The difficulty with observing Venus is caused by its thick, dense atmosphere composed of 96 percent carbon dioxide with a pressure at the surface 90 times the atmospheric pressure on Earth. For comparison, the pressure on the surface of Venus is about the same as the pressure 1 kilometer below sea level on Earth.

The dense atmosphere of Venus and its cover of reflecting sulfuric acid clouds 50 to 70 kilometers above its surface effectively hide the planet from optical observation. Instead, Venus's surface topography must be explored by radar, either from giant dish antennas on Earth or from orbiters circling Venus above its atmosphere.

The United States space probe Pioneer Venus has mapped the surface features larger than 30 kilometers across, while the more recent Russian probes, Venera 15 and 16, have made maps of features down to a horizontal scale of 1 to 2 kilometers. These radar maps have shown two major areas of relief: a northern continent (a continent in the sense of a high region; there are no oceans on Venus) with mountains up to 11 kilometers above the average surface elevation; and a smaller region, Beta

Regio, which has a large shield-shaped mountain about 500 kilometers across and 4 kilometers high named Theia Mons. The Russian radar has also mapped a 100-kilometer-wide oval depression on the northern continent, which may be a caldera.

Earth and Venus are much alike in radius and mass, but the "continents" on Venus cover only about 5 percent of its known surface in contrast to 35 percent on Earth. In addition, no features have yet been seen on Venus to suggest that plate tectonics has been an important process there. Data from Russian space probes that landed on Venus indicate the surface has a composition similar to basalt, but with a higher sulfur content. These probes also measured the traces of radioactivity from Venus rocks and found them similar to surface rocks on Earth.

The gases in Venus's atmosphere – carbon dioxide, nitrogen, water vapor, and sulfur compounds – suggest that they may be of volcanic origin. On Earth, the great bulk of steam and carbon dioxide exhaled from volcanoes over geologic time have formed oceans and limestone. Carbon dioxide dissolves in the ocean water, and is removed by algae and corals to form limestone reefs. However, on Venus, whose surface temperature is a searing 460° C, liquid water and limestone rocks cannot form, and carbon dioxide and water vapor stay in the atmosphere. It is difficult to determine cause and effect; the greenhouse effect of the high amounts of carbon dioxide in Venus's atmosphere traps the sun's radiation and causes the high surface temperature. On the other hand, it is the high surface temperature that has kept any oceans from forming on Venus, and hence prevented any opportunity to form limestone (calcium carbonate), which would remove carbon dioxide from its dense atmosphere.

Once an Earthlike planet forms either oceans or a hot, dense atmosphere, it may remain that way. Venus would be intolerable for life as we know it on Earth.

Ancient volcanoes probably did have a major role in the evolution of Venus, and it is also possible that active volcanoes still erupt on that mysterious planet (Fig. 5.5). Two lines of indirect evidence suggest that eruptions have taken place during the past decade. First, between 1978 and 1984, orbiters measured a 90 percent drop in the SO_2 gases in Venus's high atmosphere. This suggests that major amounts of SO_2 may have been injected into it by large volcanic eruptions just before 1978, and that chemical reactions with water vapor and surface materials have been steadily removing the SO_2 during a period of lesser volcanic activity. Second, both American and Russian space probes of Venus have detected low-frequency radio signals thought to be caused by lightning strikes. These strikes occur in restricted areas, one of the more conspicuous being

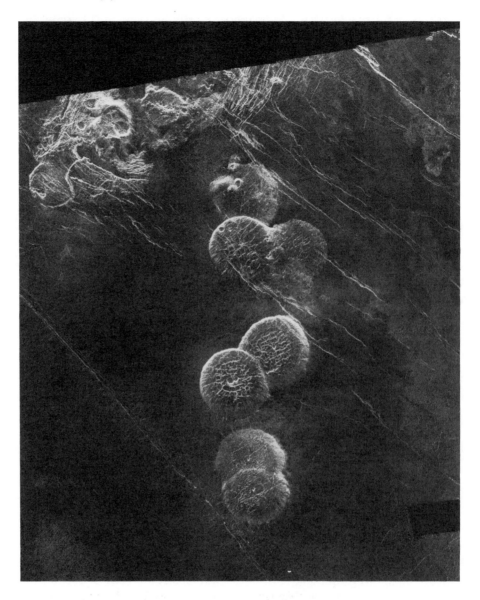

Figure 5.5. Radar images of Venus from NASA's Magellan spacecraft show many landforms that appear to be of volcanic origin. The seven dome-shaped hills in this image are each about 25 km in diameter and as much as 750 meters high. They are thought to be lava domes. (November 7, 1990, image by NASA, JPL)

Beta Regio – an area of shield-shaped mountains. Charges of static electricity build up in turbulent ash clouds, and lightning bolts accompany explosive ash cloud eruptions on Earth. However, volcanic eruptions on Venus may be less explosive because the great surface pressure of its dense atmosphere would inhibit rapid expansion of volcanic gases, the process that causes explosive eruptions on Earth.

The changing sulfur dioxide levels in Venus's high atmosphere and the static discharges are enticing clues, but there is much more to learn about the hidden face of Venus before any definite conclusions can be reached about active volcanoes on Earth's sister planet.

IO

Io, named for the mistress of Zeus, is the innermost moon of Jupiter and about the same size and mass as Earth's moon. Io also appears to be the most volcanically active body in the solar system. Images of Io taken by the Voyager space probes during their flybys in March and July 1979 have drastically changed our views about volcanic processes. Tidal friction and slow radioactive decay have been established as important sources of internal heat to generate magma, and the concept of magma itself has been extended to molten systems other than silicate rocks. Sulfur and sulfur compounds appear to play a major role in the volcanic eruptions on Io.

Color images processed from faint electronic signals created tremendous excitement among planetary scientists. These pictures revealed volcanic features dominating the entire face of Io. Eight plumes of gas and particles as much as 300 kilometers high and 1,200 kilometers wide were seen spewing into the near-vacuum of Io's atmosphere. About 200 shallow calderas more than 20 kilometers in diameter were identified in the pictures. These black calderas are surrounded by apparent flows of black, red, orange, and yellow materials – a landscape strange and colorful beyond imagination. The interpretation that some of the calderas on Io may contain active lava lakes is supported by evidence of color-pattern changes within three of the calderas during the four months between Voyager encounters.

One major question about volcanism on Io is the character of the magma: whether it is molten sulfur and sulfur compounds, molten silicate rock as on Earth, or some combination of both. David Pieri, a scientist with the Jet Propulsion Laboratory, points out that sulfur has been identified as an element in the thin atmosphere and space near Io, and that the surface colors of apparent flows radiating from Ionian calderas – from black through red to orange are the same as molten sulfur as it cools (black > 221° C; red 171–221° C; red-orange 161–71° C; and orange < 161° C). Susan Kieffer of the U.S. Geological Survey has calculated that both sulfur dioxide and sulfur could erupt from subsurface reservoirs, and that temperatures required to form a sulfur plume would probably have to exceed 427° C. Michael Carr, also of the U.S. Geological Survey, argues that sulfur does not have the strength to form the steeper relief features seen on Io. Some calderas are deeper than 1 kilometer, and a few steep mountains rise as much as 10 kilometers. He proposes that part, or even most, of the volcanic rock on Io is basaltic.

Although the exact composition of Ionian magmas remains a problem for future exploration, it is clear that Io generates and releases an extraordinary amount of internal heat – about 1.0 to 1.5 watts per square meter, some 20 times the internal heat flow from Earth. This volcanic energy is maintained by the continued flexing of Io's surface layers as they are pulled first one way by the immense gravity of Jupiter and then another by the tug from Europa, the next closest Jovian moon.

The exploration of space in the past few decades has changed and broadened our views of geologic processes in revolutionary ways. Comparison of features and processes among the planets and moons is not only startling but also highly educational. What we take for granted on Earth may be only a special case of a much more general concept.

COMPARATIVE VOLCANOLOGY

Volcanic activity on the Earth, Moon, Mars, Venus, and Io has interesting similarities and differences. On the first four bodies, silicate rock volcanism, mainly basaltic, has continued over time spans proportional to the diameter of the planet or moon. On the Moon (3,476 kilometers in diameter), the major period of volcanic eruptions ceased about 3 billion years ago; on Mars (6,786 kilometers), about 200 million years ago; and on Venus (12,100 kilometers), as on Earth (12,756 kilometers), volcanoes are apparently still active. Heat was and still is being generated in each planetary body at a rate proportional to its volume; that is, proportional to the cube of its radius. For example, if all other factors were equal, Mars, whose radius is about two times that of the Moon, should generate eight times more internal heat. Because heat is lost through the surface area of a planet or moon, the loss should be roughly proportional to the square of its radius. In this example, Mars should lose heat four times as fast as the Moon. Generation and loss of internal heat both increase with size, but generation increases faster than loss. Planets, like living animals, are heat engines, and the big ones stay warmer longer than the small ones.

The explosiveness of volcanoes on various planets and moons is controlled by both internal and external factors. High gas content in viscous magma promotes explosive boiling, while dense, heavy atmospheres tend to inhibit rapid boiling. If Mount St. Helens had exploded in the vacuum of the Moon or in the low-density atmosphere of Mars (7 percent of the density of Earth's atmosphere), the swath of destruction would have been much greater and would have extended in all directions. The fact that most of the volcanic activity on the Moon and Mars has built lava plains or shield volcanoes, generally a sign of effusive action, strongly argues that their magmas were low in gas content and low in viscosity. The rocks brought back from the Moon are dense with almost no bubble holes, confirming this deduction.

If it had occurred on Venus, the Mount St. Helens explosion would have been much less violent because of the retarding effect of Venus's dense atmosphere. That sulfuric acid clouds and lightning bolts seem to be associated with some explosive volcanic eruptions on Venus, despite the inhibiting high-pressure atmosphere, implies that some magma there is both highly viscous and highly charged with gas. Once the general processes are known, the mind's eye can probe into places that direct visual observations may never reach.

Io has been eye-opening. It is only the size of the Moon, but for its size it has 20 times the volcanic activity of Earth. It points out two major new ideas about volcanic processes in the solar system: that the internal heat engines in multiple moons may be driven by tidal friction, and that other magma systems besides silicate rocks may erupt and build volcanolike features.

Io, however, does conform to the generality that explosive eruptions on planets or moons with low-density atmospheres can be extremely violent. The atmospheric pressure on Io is nearly a vacuum, and the explosive plumes spray as high as 300 kilometers above their vents. Susan Kieffer points out that these eruption plumes may be more analogous to a geyser eruption than to a volcanic eruption. She calculates that an eruption of Old Faithful Geyser on a moon with low gravity and low atmospheric pressure like Io would rise to 38 kilometers, more than a thousand times its normal 30-meter height on Earth.

Part I has explored the physical setting and processes of volcanoes, on Earth and in the solar system. Part II takes a closer look at the character and origin of volcanic rocks and gases.

PART II

Volcanic rocks

The insignificant and the extraordinary are both the architects of the natural world.

– Carl Sagan

6

Molten rock

Johannesburg, South Africa: 1950

Photojournalist Margaret Bourke-White, in South Africa on assignment for *Life* magazine, watched with delight as a group of young gold miners celebrated their Sunday off with a performance of tribal dances near a mine entrance. Their liveliness and grace was so appealing, she decided to use two of the dancers as subjects of a photo-essay, starting with the dance and then following them through a normal work day. When she talked to the mine superintendent to make arrangements, he cautioned her that the two black miners – identified by tattooed numbers instead of names – were assigned to work in an especially deep and hazardous part of the mine, but she persisted and was finally granted permission to descend to the "remnant area" where visitors were never taken.

The next morning she reported to the Robinson Deep, one of the oldest mines in the region; a mine that plunged to a depth of 2 miles beneath the modern city of Johannesburg. As she described the experience,

> I was told ... to put on proper mine clothes. I was astonished to find that these were heavy and warm, as if I were dressing for the frigid top of the world rather than for the hot depths of the earth. The superintendent explained to me that unless the body is thoroughly conditioned to the abrupt temperature changes under the earth's crust, a visitor might go through the heat of the lower depths, get wringing wet and catch pneumonia on the slow return to normal temperatures aboveground. My costume was topped with a crash helmet, and I wore a whistle hanging round my neck to use if we were trapped.
>
> It was a solemn moment when I stepped into the mine cage and started the slow two-mile descent into the hidden space of the world. I felt something of the excitement of my first snorkeling in tropical waters: my first look into a new world. With snorkeling, all was spar-

kling and bright; here it was all gloom and obscurity. As the elevator lumbered downward, the darkness was broken only by occasional eerie cracks of light as we passed the mine stages.

At the bottom of the first mile, which is the halfway point in this vertical journey, you change to a smaller mine car which shoots down an incline. You travel that second downward mile in the sober realization that you are now below sea level.

When I stepped out, I felt no discomfort, for the air circulation system close to the elevator shafts was adequate. As soon as we started walking along the lateral passages, the atmosphere became very hot and humid. When we reached the little sloping pocket where the two men were working, I could hardly recognize my dancers. With rivers of sweat pouring down their bare chests, and with sad eyes and perspiration-beaded faces, they hacked away. I was in the midst of making photographs when a strange depression and lassitude overcame me. I could hardly raise my hands; I had lost the power of speech.

The superintendent, noticing my distress, led me quickly to a more open mine-passage, gave me a little water, directed me to wash out my mouth, but not swallow it; then he took me to the foot of the shaft, where I was revived by the better air. Later I learned a man had died from heat prostration two months previously on this mine face.*

THE EARTH'S HEAT

The oppressive heat in a deep mine is always a serious problem for mine operators. As high as the temperatures were in the South African mine described by Margaret Bourke-White, some mines – especially in areas of recent volcanism – are even hotter.

The old silver mines of Nevada's Comstock Lode were unusually hot; at depths of 1,500 to 2,000 feet the rock was so hot, it was painful to touch, and crevices or drill holes sometimes gushed with streams of boiling water. Air pipes and blowers were used extensively, not only for fresh air for the miners to breathe, but to cool off the rock and help keep the heat down. Even so, miners often worked in temperatures as high as 110° F. It was not uncommon for a miner – or visitor – on reaching the cool air at the surface after spending hours in the intense heat to faint and fall down the shaft to his death from the mine car.

The increase of temperatures with increasing depth in the Earth is known as the "geothermal gradient." In addition to the evidence from mine shafts, it has now been established by deep drill holes and by special

---

* Margaret Bourke-White, *Portrait of Myself* (New York: Simon & Schuster, 1963).

temperature probes pushed into soft sediments on ocean floors that the gradient is a worldwide phenomenon.

The average geothermal gradient is about 2° to 3° C per 100 meters of depth, indicating that heat is flowing outward from inside the Earth. The average heat flow amounts to 0.06 watts of power for each square meter of the Earth's surface. Although this is a tiny amount compared to solar power, which averages about 40.0 watts per square meter of Earth's surface, on a worldwide basis it is an enormous quantity of heat. The thermal power loss from the Earth is more than double all the power presently produced by the burning of wood, coal, oil, and gas.

The source of Earth's internal heat has long been a topic of investigation. In the late nineteenth century, British scientist Lord Kelvin attempted to calculate the age of the Earth from its cooling rate. He conjectured that if the Earth had been created as a molten globe, it would have cooled to its present state in 20 million to 40 million years. He placed Earth's age at less than 50 million years; he was wrong by more than 4 billion years. What was missing from his calculation was the effect of radioactivity, which is an important component of heat production within the Earth.

Since the discovery of radioactivity in 1895, scientists have found that naturally occurring radioactive elements within the Earth – such as uranium, thorium, and potassium – slowly disintegrate, producing small but steady amounts of heat. Rocks are excellent insulators, and thus heat slowly accumulates over tens of millions of years. Resultant temperatures are high enough to partially melt the rocks in which the heat is generated.

If the increase in temperature with depth in the Earth maintained a steady rate, the 1,000° C melting temperature of rock would be reached at depths of about 40 kilometers beneath the surface. However, most of the heat is being generated by the radioactive elements in rocks near the surface, so the rate of temperature rise decreases with each additional kilometer of depth (Fig. 6.1).

Some of Earth's interior heat may also be left over from the formation of our planet by collisions of material during the violent birth of the solar system. High-velocity impacts of material pulled in by the gravity field of the growing Earth would have released enormous amounts of heat. Depending on the rate at which the collisions occurred, the Earth may have formed in a partly molten state. It has also been calculated that the gravitational energy released as the common heavy element iron settled toward the center of the Earth would generate enough energy to melt most solid rocks left over from the collision and accretion stage.

A third possible source of continuing heat generation inside the Earth is tidal friction. The energy that raises tides in the ocean is supplied by

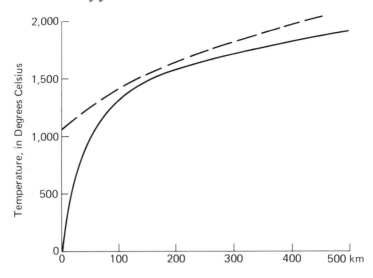

Figure 6.1. Graph showing the increase in temperature in the Earth (solid line) and melting temperature of basalt (dashed line) with increasing depth in the Earth. These data are approximate because geothermal gradient–the slope of the temperature line–differs in various geologic settings. Nevertheless, the trends of the lines are correct and demonstrate some important concepts: (1) The geothermal gradient becomes less with increasing depth because the common radioactive elements–uranium, thorium, and potassium–that contribute most to the internal heating of the Earth are more abundant in rocks within the first several kilometers beneath the surface. (2) The close approach, and sometimes intersection, of the two lines between about 100 and 200 km depth is the zone of partial melting in the Earth's upper mantle.

the slow decrease in the immense amount of energy stored in the rotation of the Earth–Moon and Earth–Sun systems. Most tidal energy is used in the movement of seawater, but some fraction of tidal energy may bend and unbend rock layers deep within the Earth, releasing heat in a manner analogous to bending a wire back and forth until it feels hot to the touch.

MAGMA AND THE EARTH'S INTERIOR

Studies of earthquake waves and the speeds at which they travel through the Earth indicate there is a solid metallic inner core, a molten metallic outer core, a solid rocky mantle, probably with a zone of partial melt within its upper region, and a solid rocky crust on top (Fig. 6.2).

Temperatures high enough to melt rocks and produce magma probably occur in the upper mantle at depths of 70 to 200 kilometers. That is the root zone of volcanoes. Some minerals melt at lower temperatures than others, so the magma formed is a mush of hot, viscous liquid

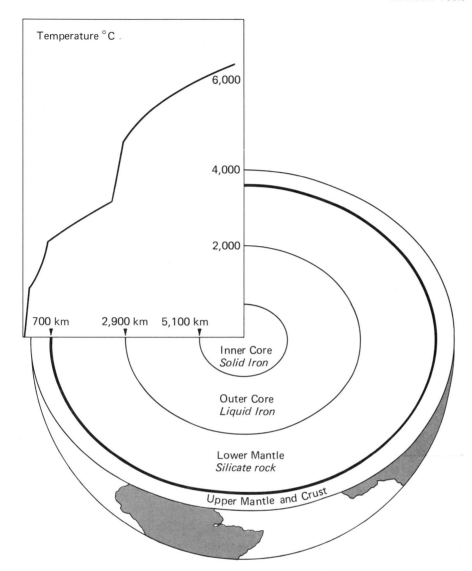

Figure 6.2. Estimates of temperature inside the Earth have recently been revised upward. The temperature at which iron melts is the key to the actual temperature of the boundary of the solid inner core and the molten outer core. This boundary occurs at a depth of 5,100 km, according to analysis of earthquake waves traveling through the Earth's interior. New determinations of the melting point of iron under the high pressure at that depth exceed 6,000° C (older estimates were about 4,000° C). (After Q. Williams et al., Science 236, *no. 4798 [1987]: 181)*

surrounding hot but still solid crystals. Geologists refer to this as a "zone of partial melt." Until the melt fraction exceeds about one-half, the mix of crystals and melt behaves more like a solid than a liquid.

Most magma formed by partial melting in the root zones of volcanoes is basaltic in composition. It is composed of about 50 percent silica (SiO_2); 15 percent alumina (Al_2O_3); and lesser amounts of iron, calcium, and magnesium. Basalt magma is lighter than the crystals in the zone of partial melting and tends to rise by buoyancy and form large pockets of molten rock called "magma chambers." The temperature of basalt magma is about 1,200° C, and if it begins to cool, olivine crystals will form from the melt. The mineral olivine is high in magnesium and iron, and low in silica compared to basalt. As the heavier olivine crystals settle downward toward the bottom of the magma chamber they leave the still-molten magma enriched in silica and depleted in magnesium and iron. This is one major process by which magma can change its chemical composition.

Magma also changes composition by partly melting the host rocks through which it ascends. Basalt magma at 1,200° C has enough heat to melt some rocks and minerals with relatively low melting temperatures when it comes in contact with them. Because the crustal rocks beneath continents tend to be higher in silica, sodium, and potassium than basalt magma, partial melting and assimilation of some of these host rocks will alter the composition of the primary magma. This is particularly true if the melted host rock does not mix thoroughly with the basalt magma. An analogy would be using hot water to melt wax. This would form two separate liquids – hot wax floating on hot water. Although magma batches probably mix better than water and wax, there is evidence that silica-rich magma can "float" on top of basaltic magma in reservoirs of molten rock beneath some volcanoes in continental settings.

This can be graphically demonstrated by studying deposits from ancient volcanic eruptions. Good examples are seen in the ash-flow deposits around Mount Mazama in the Cascade Range of the northwestern United States, which erupted about 7,000 years ago and formed the caldera that holds Crater Lake. Those deposits are silica-rich at the base (the part that would have erupted first from the top of the magma chamber), and silica-poor near the top (the magma erupted later from deeper in the magma chamber (Fig. 6.3).

Volcanic rocks that are formed when magma is erupted to the surface are classified into major clans – basalt, andesite, dacite, and rhyolite. Average silica content of typical rocks in each clan are, respectively, 50 percent, 55 percent, 65 percent, and 73 percent. Silica content is not the only characteristic determining volcanic rock type, but it is one of the most important.

The physical properties of magma are also related to the amount of

Figure 6.3. The pyroclastic flows from the caldera-forming eruption of Crater Lake, Oregon, which occurred nearly 7,000 years ago, have been eroded along stream canyons into pinnacles. Notice how the lower portion of the deposit is lighter colored (silica-rich) and the top part is darker (containing less silica). Since the top of the magma chamber was erupted first, this canyon provides an inverted picture of the changing composition of the magma chamber.

silica it contains. Basaltic magma is more fluid than silica-rich dacitic magma. The viscosity of Hawaiian basaltic magma is about that of honey at room temperature, while dacitic magma, like that slowly extruded in the lava dome at Mount St. Helens, is more like tar.

VOLCANIC GASES

Magma has substantial amounts of volcanic gas dissolved in it, in the same way that soft drinks or champagne contain carbon dioxide (CO_2) gas dissolved in them under pressure. When magma reaches the Earth's surface, much of its dissolved gases – carbon dioxide, steam, and sulfur gases – are released. This can occur explosively if the pressure is suddenly reduced, like the cork popping from the champagne bottle, or less vi-

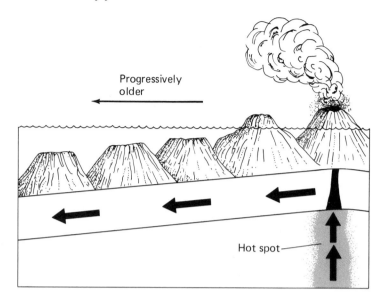

Figure 6.4. Cartoon showing a ridge of progressively older volcanoes like the Hawaiian chain, formed by a plate moving over a hot spot. The extinct volcanoes are eroded to nearly flat surfaces at sea level and are then submerged as flat-topped seamounts called "guyots." (After Charles Plummer and David McGeary)

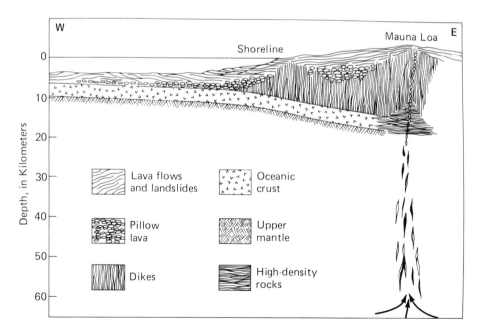

Figure 6.5. Schematic cross section of Mauna Loa Volcano, Hawaii. Basaltic magma from a source below 60 km rises almost continuously through conduits into a shallow magma chamber 3 to 6 km beneath a summit caldera. (After Dave Hill, John Zucca, and Jerry Eaton, U.S. Geological Survey)

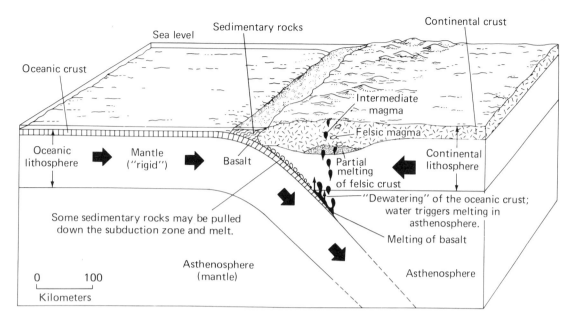

Figure 6.6. Cartoon showing the nature of volcanoes formed at subduction zones. The asthenosphere is the zone in the upper mantle that closely approaches or reaches the melting temperature of basalt; it behaves in a plastic, low-strength manner. (After Charles Plummer and David McGeary)

olently if pressure is reduced gradually so that the gases can slowly leak away.

Volcanic gases in Hawaiian type volcanoes constitute about 0.5 percent of the total weight of the magma. This seems like a small amount, but at magmatic temperatures and low pressure at the Earth's surface, the gas boiling out of Hawaiian magma expands to hundreds of times the volume of the molten rock in which it was dissolved under pressure.

PHYSICAL PROPERTIES OF MAGMA

Magma rises by buoyancy. The density of basaltic magma is about 2.7 grams per cubic centimeter, and more silica-rich magma is less dense. The rocks in the lower crust and upper mantle have densities near 3.0 grams per cubic centimeter. This difference in density causes magma to rise. The ascent of magma is retarded by its viscosity, its yield strength, and the strength of the rocks it must penetrate on its way toward the surface.

In general the viscosity of magma increases as the composition changes from basalt through andesite and dacite to rhyolite. As a result, basaltic

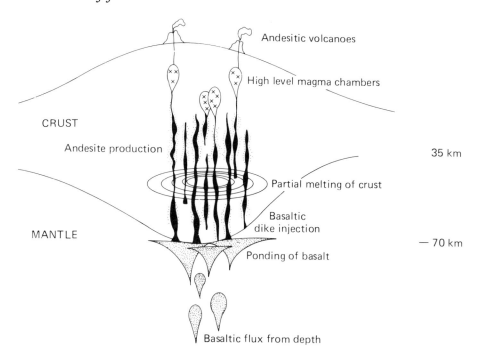

Figure 6.7. Schematic cross section of andesitic stratovolcanoes. Basaltic magma rises from a zone above a subducted plate, possibly 100 to 150 km down. As it reaches the more brittle rocks in the Earth's crust, it ascends through multiple dikes toward the surface. As the magma ascends through the crust contamination with crustal rocks leads to an increase in silica. Contamination and crystal fractionation in the high-level-magma chambers may increase the silica content even further. (From S. L. de Silva, "Antiplano-Puna Volcanic Complex of the Central Andes," Geology 17 [1989]: 1105)

volcanoes tend to be effusive and erupt quite frequently in comparison to the more silica-rich magmas which erupt more explosively and less often, as explained in Chapter 3.

Magma behaves more like a plastic than a fluid; that is, it has a yield strength that must be overcome before it will begin to flow. Toothpaste is an example of a common substance with a yield strength. Hold an open tube upside down and no toothpaste comes out – its yield strength resists the force of gravity; but when the tube is squeezed the yield strength is easily overcome and the toothpaste flows out.

Both viscosity and yield strength of magma change with temperature. Lowering the temperature increases both; as magma cools it loses its plasticity and becomes solid. When magma solidifies underground without erupting to the surface, it is called an intrusive igneous rock; for example, rhyolitic magma becomes granite. If magma erupts in a fluid state, it is called lava, but if it cools rapidly during eruption and is dispersed into hot solid fragments it forms pyroclastic deposits.

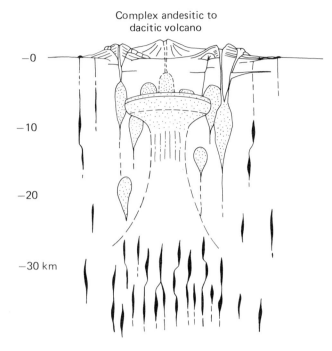

Figure 6.8. Schematic cross section of a complex andesitic–dacitic volcano. As the life-span of an andesitic stratovolcano (see Figure 6.7) increases, more partial melting of host rocks takes place in the lower crust. This silica-rich magma intermittently rises in blobs toward the surface. Some of these upward intrusions coalesce into a stratified magma chamber beneath the central cone. Surface features include stratovolcanoes and lava domes. (After Wes Hildreth, "Gradients in Silicic Magma Chambers," Journal of Geophysical Research *86, no. B11 [1981]: 10179)*

DEEP INSIDE VOLCANOES

The deep underground structure of volcanoes remains a mystery, but its study is being approached in several ways. In some parts of the world, such as Scotland or Nevada, erosion has exposed the insides and shallow roots of volcanoes that have been dead for millions of years. Examining the structure of such exhumed volcanoes helps us understand how the magma reached the surface, even though these ancient volcanoes only show the final stages of their underground structure – not how they looked when the volcanoes were vigorously active.

Studying the pattern of earthquakes beneath active volcanoes is also useful in trying to understand their deeper structure. The location of the earthquakes and the way in which the seismic waves pass through the rocks beneath active volcanoes provide a hazy but useful picture of the subsurface structure. Seismographs serve volcanologists as X-ray machines serve doctors, but the "picture" beneath volcanoes is much less clear.

Using all the clues they can find, some geologists have drawn schematic cross sections of how they believe active volcanoes appear at depth. The models illustrated here show how magma may rise and accumulate beneath a typical Hawaiian shield volcano (Figs. 6.4 and 6.5), and also show a typical andesitic stratovolcano and a more complex volcano erupting dacitic magma (Figs. 6.6–6.8). These models of how volcanoes work are plausible, and are the best picture available at present. As new facts and concepts are discovered, such models will undoubtedly be modified. As semanticist S. I. Hayakawa said, "The map is not the territory."

7

Lava flows

Catania, Sicily: March, 1669

Ominous earthquakes shook Sicily's Mount Etna for three days, giving warning to the villagers who lived on its slopes of terrors yet to come. Etna's history had been long and violent, but the series of eruptions that began on March 11 were the worst they had ever experienced or, indeed, imagined.

On March 25, under a sky still dark with ash from the volcano's explosions, those living on the south side of the mountain could see the terrifying glow of a surging stream of lava flowing downhill toward them. Its advance was relentless; demolishing several small villages on its way, it headed for the ancient city of Catania, which lay 16 kilometers from Etna's summit.

While most Catanians were gathering their belongings to flee from this inexorable natural force, one group decided to stay and fight. Diego de Pappalardo marshaled 50 strong men, and together they planned a strategy to save the village. First they soaked cowhides in water to wrap themselves in, as protection against the overpowering heat of the molten lava; then they found some long iron bars, and headed up the mountain.

By the time they reached the surging flow it had established a central channel; where the edges had cooled, the lava had built up into high levees with the molten stream flowing between them. Shielded by the wet cowhides, the Catanians were able to hack out a gap in one of the levees through which a tongue of lava poured, greatly slowing the main stream. Their problem now was to keep the breach open.

While they were celebrating their initial success, though, some citizens of the nearby town of Paterno noticed that the newly diverted flow was now moving directly toward their town. The Paternans in turn mobilized and 500 of them charged up the hill, forcing the Catanians away from

Figure 7.1. Lava diversion embankments built to help protect the Mauna Loa Atmospheric Observatory. The barrier's V shape pointing uphill is designed to deflect descending flows to one side or the other of the observatory. (Photograph by J. P. Lockwood, U.S. Geological Survey)

the lava levees. The breach soon healed; the flow resumed its original course, eventually engulfing much of Catania.

CONTROLLING LAVA FLOWS

More than three centuries later, attempting to control lava flows is still controversial. It has been tried in several locations with varying degrees

Figure 7.2. In 1973 a thick, blocky lava flow from a new volcanic vent on Heimaey, an island off the south coast of Iceland, threatened to cut off the entrance to the harbor. Because fishing is the principal industry on Heimaey, a major effort was made to stop the flow; enormous volumes of seawater were pumped onto the lava to cool it and slow its progress. This appeared to work, and combined with the good fortune that the eruption stopped, the harbor was saved. (Photograph by James G. Moore, U.S. Geological Survey)

of success. Two approaches have been used: building barriers to divert a flow around a populated area or buildings (Fig. 7.1), and altering the direction of a flow or slowing it by disturbing its margins. Barriers erected on Etna during a 1983 eruption were partially successful, although some were overwhelmed by later phases of the eruption. Diversion barriers have been built above the Mauna Loa Weather Observatory in Hawaii, but their effectiveness has not yet been tested. Slowing an advancing flow by cooling its margins has been tried with some success, most notably at Vestmannaeyjar in Iceland. There, in 1973, large amounts of seawater were continuously sprayed on the advancing front of a flow

Figure 7.3. Ropy surface on a pahoehoe lava flow formed during the Mauna Ulu eruption of Kilauea Volcano, Hawaii, in 1973. The flow moved from left to right and the surface wrinkled into ropelike coils as additional lava pushed in from the left.

that threatened to close the harbor. Such action apparently slowed the flow, but it is not clear whether it influenced its final length (Fig. 7.2).

In all attempts at lava flow diversion, the political questions are as thorny as the technical ones. For instance, bombing the feeding channel of an active flow to alter its course is a technique often suggested, and it was tried with arguable results in Hawaii in 1935. The idea is usually enthusiastically endorsed by those living in the path of the flow, but understandably resisted by landowners on either side. No one is anxious to test the problem of who is financially responsible for damage done by a lava flow that has been influenced by man and thus can no longer be considered an "act of God."

AA AND PAHOEHOE

Perhaps one answer to lessening the danger from lava flows is in understanding the character of the flows themselves: how they differ from one another, how they can be expected to behave, and of what material they are composed. Not all volcanoes produce flows of fluid lava; those

Figure 7.4. Pahoehoe (left) *and aa* (right) *lavas on Kilauea Volcano, Hawaii. These flows, formed in 1972 during the Mauna Ulu eruption, are similar in composition but different in physical properties and appearance.*

that erupt explosively may produce mostly ash, bombs, and hot fragments sometimes emulsified by gases. Among the volcanoes that erupt effusively the two major kinds of lava flows are called "pahoehoe" and "aa," Hawaiian words that have been adopted worldwide to describe those characteristic types. Pahoehoe refers to a lava flow with smooth to ropy surfaces (Fig. 7.3), while aa refers to flows whose surfaces are covered by thick, jumbled piles of loose, sharp lava blocks (Fig. 7.4). Ancient Hawaiian trails across barren flows were built on pahoehoe wherever possible.

Pahoehoe is usually produced in high-temperature (low viscosity) eruptions, with low-volume rates of emission. High lava fountains that cool the lava clots before they land and re-form into flows, high-volume rates of emission, and steep slopes that speed the movement of the flow all tend to produce aa. Fudge is a fair kitchen analogy of these two kinds of lava flow. If it is poured hot and allowed to cool into a smooth slab, it is like pahoehoe; if it is stirred too much and cools before pouring, it "sugars," or crystallizes into a rough broken slab that crumbles to pieces. A pahoehoe flow sometimes transforms into aa, but the reverse is rare.

Watching a pahoehoe lava flow creep and sizzle across a lawn toward a doomed house is like watching a huge snake slowly and relentlessly approach its prey. The orange to black flow about 30 centimeters thick moves ahead in a blunt wedge at a speed of roughly 1 meter per minute. The cooling skin on the top of the flow is stretched by the molten basaltic lava as it inflates the advancing lobe. The tough, glassy skin becomes thinner as it stretches, and orange blotches of more incandescent melt appear on the spreading surface.

As the flow inches forward its movement resembles the caterpillar track on a bulldozer. The advancing top surface is rolled over and under the front of the moving lobe. Individual lobes creep forward a few meters and then stop as other lobes from the advancing flow front catch up. The overall front may be several hundred meters across, whereas individual lobes in which the flow motion is concentrated are generally only a few meters wide.

Naturally there are variations in the speed and character of the advance of a pahoehoe flow. Although most flow fronts creep ahead at speeds of less than a meter per minute, an unusual pahoehoe flow produced at high volume on a steep slope can move at speeds as high as 400 meters per minute, so fast that it would be difficult for a person to outrun it.

Upstream in an active pahoehoe flow much of the surface hardens into smooth black patches interrupted by wrinkled areas that look like coils of rope. A pile of black candle drippings magnified about 100 times would have a roughly similar appearance. Beneath the hardened surface of a pahoehoe flow, hot lava still flows rapidly in tunnels that supply the advancing flow front. These lava tubes form a complex network that transports molten rock from the vent to flow front over distances of several kilometers (Fig. 7.5). Individual flow lobes of pahoehoe are generally less than a meter thick, but as they pile on top on one another, the overall flow may become several meters thick. Small lava tubes feeding individual lobes apparently coalesce as the multiple flow lobes pile up, and form lava tunnels as much as 10 meters in diameter and many kilometers long.

Figure 7.5. Thurston Lava Tube in Hawaii is about 3 to 5 meters in diameter and more than 100 meters long. Lava tubes are formed when pahoehoe flows crust over and molten lava continues to move downslope beneath the crust. When the eruption stops, the molten lava in the tube drains away, leaving an empty tunnel. (Photograph by Jane Takahashi, U.S. Geological Survey)

The roofs of these active lava tunnels sometimes collapse, or clots of semihardened lava may clog part of the flowing system. These disruptions in the lava feeding system cause the advancing flow front to slow or halt, as breakouts on the sides or top of the hardened upstream flows interrupt the transport in the tunnel network. When this happens, the volume of molten lava in tubes downstream from the blockage will lessen or the tubes may even empty completely. Days or weeks later new molten rock may find its way back into the old clogged tunnel system, or a new network of tunnels may evolve and change the entire course of the lower part of the flow field.

Pahoehoe flows from long-lived eruptions lasting several months or years can overplate many square kilometers of land with accumulated thicknesses of tens of meters. In Hawaii, these flows often reach the sea, adding new land to the still-growing island.

Although aa flows are the same composition as pahoehoe flows, their physical characteristics are very distinct. Watching the advance of an aa flow is a much different experience from observing the quietly creeping front of a pahoehoe flow. An aa flow front is a steep-sloping wall of jumbled dull red lava lumps that can crash through a forest, pushing down large trees instead of flowing around them. An individual aa flow is about 2 to 5 meters thick and consists of a molten core with about 0.5 to 1 meter of broken but still hot lava blocks riding on top. As this rubble reaches the flow front it tumbles down the steep advancing wall and is overrun by the flow. When seen in a road cut, an old aa flow has three layers; rubble on top, a massive core of solid rock, and a jumble of more lava blocks on the bottom.

The front of an aa flow often advances in surges, with the flow front gradually growing higher before a surge begins. The flow front may move only a few meters downhill in an hour, but then surge forward a hundred meters in a few minutes, diminishing to its original thickness by the time the surge stops. The advancing front of a typical aa flow is about 100 meters wide, but there can be large variations in velocity and dimensions. Upstream, these flows are fed by an open channel of incandescent lava that moves at speeds of 10 to 20 kilometers per hour. This river of lava is generally about 10 meters wide – quite narrow compared to the overall width of the black, rubble-covered aa flow – and it breaks up into many slower-moving distributary channels as it approaches the flow front.

An aa flow does not usually extend as far from its vent as a pahoehoe flow does. The basic reason for this is the nature of the open channel system that feeds an aa flow; it loses heat more rapidly than the sealed-over lava tubes feeding a long-lived pahoehoe flow. When an aa flow does reach the ocean it can generate sizable steam explosions because of its large, rough surface and generally high flow volume coming in contact with the seawater; sometimes it builds a small cone of lava fragments at the shoreline. In contrast, the smaller and smoother pahoehoe flows enter the sea with remarkably little agitation.

About 99 percent of the island of Hawaii is composed of aa and pahoehoe flows; the other 1 percent is made up of cinders and spatter near vents and the ash from rare explosive eruptions. The flows cool in a matter of weeks or months, depending on the thickness of the individual flow, and then their barren surfaces begin to weather into soil. The speed of this weathering is very dependent on rainfall and temperature. Where the climate is warm and wet, plants begin to grow on new lava flows within a few years and the vegetation itself helps speed the rock decomposition. Flows in humid tropical areas can be hidden by trees and grass in 50 years. In contrast, in dry areas, flows 500 years old or more appear almost as barren as when they were formed.

Figure 7.6. Aerial view of a prehistoric block lava flow from Mount Shasta, California. U.S. Highway 95 skirts a toe of the 25-meter-thick flow in the foreground. (Photograph by U.S. Geological Survey)

MORE VISCOUS FLOWS

Most lavas from continental volcanoes are more siliceous and therefore more viscous, so their flows generally differ from Hawaiian types. Block lava flows often form on andesitic stratovolcanoes. They are similar to aa flows but are thicker, much slower moving, and usually have a layer of large, partially cooled lava blocks riding on the surface of the flow hiding most of the interior incandescence of the molten, slowly moving core. As the flow inches imperceptibly forward, large rockfalls avalanche down from its high, steep front wall, sending up billowing dust clouds. Except for this action and the fact that a few meters of forward creep a day can be measured, it would be easy for an observer to believe that

Figure 7.7. The lava dome inside the crater at Mount St. Helens in June 1984. At that time the dome was 230 meters high and more than 800 meters in diameter. (Photograph by Lyn Topinka, U.S. Geological Survey)

the flow was dead. Block lava flows are often as thick as 30 meters, hundreds of meters wide, and many kilometers long (Fig. 7.6). Their sides and front form steep slopes and their surface is covered by angular blocks of lava piled into thousands of hummocks and hollows.

Lava flows of dacite and rhyolite are even more viscous and sluggish than andesitic flows. They generally form after an explosive eruption has released the more gas-rich magma as ashfalls and ash flows (described in Chapters 8 and 9). Ancient rhyolite flows as thick as 250 meters have been mapped and described in Nevada.

LAVA DOMES

In its most sluggish form, highly viscous lava sometimes piles up over a vent into a lava dome. The toothpaste analogy is useful here again: Imagine putting the open mouth of a toothpaste tube up through a snug-fitting hole in a piece of cardboard. Hold the cardboard level with the tube underneath and squeeze; the result is a mini-ture model of a dacitic lava dome. Squeeze intermittently and the model dome will grow in episodes similar to the lava dome growth that has taken place at Mount St. Helens since the explosive eruptions there in 1980.

A real lava dome has great crags of hardened lava and an apron of talus blocks on and around it (Fig. 7.7). As viscous lava is slowly extruded it cools into huge angular blocks that cover the surface. Incandescent rock is seldom seen on an active lava dome except in deep cracks or when an extensive rock fall exposes the glowing interior. As lava continues to inflate the interior, the sides of the growing lava dome become steeper, and large lava blocks on these oversteepened slopes tumble onto the talus apron surrounding the dome.

The eruptions at Mount St. Helens following the great explosion of May 18, 1980, provide an excellent example of the landforms, structures, and dynamics of lava domes. During June 1980, dacite lava began to well up into the volcano's new interior crater and formed a broad, low dome about 300 meters wide and 65 meters high. This dome was blown out by an explosive eruption on July 22. Another gas-rich eruption occurred on August 7, followed by the growth of a new lava dome on August 8 and 9. This dome was destroyed by five ash explosions on October 16 to 18 and was replaced by an even larger dome that extruded into the inner crater during October 18 and 19, 1980.

Since that time the dacite lava dome at Mount St. Helens has continued to enlarge in intermittent eruptions called "dome growth episodes" (Fig. 7.8). Most lasted from about a day to a week; one in 1983–4, however, continued for nearly a year. These episodes involved both extrusion of viscous lava on the surface of the dome and inflation of the dome by the injection of magma from within.

By 1987 the lava dome at Mount St. Helens had reached a height of 250 meters and a diameter of 1,000 meters. It still looked small compared to the vast crater formed by the May 18, 1980, catastrophe. Imagining the 155-meter-high Washington Monument completely hidden inside the new lava dome helps us comprehend its immense size.

The great diversity of lava flows results from many factors, but viscosity of the molten rock is most important. As explained in Chapter 3, viscosity of molten lava is strongly influenced by composition. Most pahoehoe flows are formed by low-viscosity basalts; and most lava domes are formed by high-viscosity dacites (Fig. 7.9).

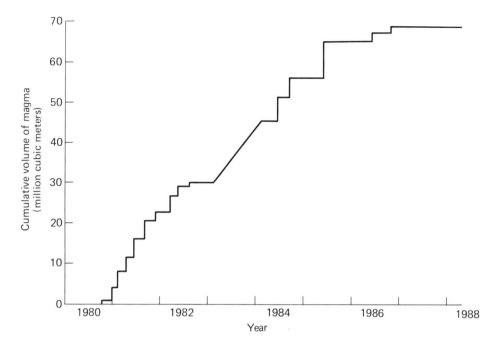

Figure 7.8. Episodes of lava dome growth at Mount St. Helens indicate a gradual decline in the magma supply rate since 1980. Most of the eruptive episodes lasted only a few days and added 1 million to 7 million cubic meters of dacitic lava to the dome. In 1983, however, growth was continuous for 12 months and totaled 15 million cubic meters. (Data from D. Swanson et al., Geological Society of America, Special Paper 212 [1987]: 3; S. Brantley, Earthquakes and Volcanoes 18, no. 5 [1986]: 217; and Smithsonian Scientific Event Alert Network Bulletins)

LAVA COMPOSITION

The exact composition of lava can be determined only by chemical analysis and microscopic identification of its constituent minerals, but field identification of common lava rock types is useful. The color, texture, and kinds of mineral grains in a lava rock specimen are all good clues to its tentative identification. In general, the color of a freshly broken surface of basalt is dark gray; andesite, light gray; dacite, light gray to light brown; and rhyolite, light tan to pink. There are many exceptions, however, especially when the lava has been exposed to acid volcanic fumes while cooling.

The texture of a lava rock refers to the size and distribution of the minerals and gas-bubble holes within the rock; it is generally controlled by the magma composition and the rate at which the lava cooled. Magma

Figure 7.9. Growing lava dome in the boiling crater lake of La Soufrière Volcano, on the island of Saint Vincent in the Caribbean. The lake, about 1 km across, was 175 meters deep at its center before the dome began to form in October 1971. (Photograph taken on December 13, 1971, by Haraldur Sigurdsson)

becomes molten lava by definition when it erupts, but the magma may have contained crystals of silicate minerals surrounded by molten glass at the time it issued forth.

The word *glass* as used here needs some explanation. The general notion is that glass is a transparent, brittle material manufactured for windows and bottles, but in fact it is a common natural material. By definition, glass is produced by melting, and has cooled to a rigid condition without forming crystals. It may be colorless or colored, translucent or opaque, through the presence of dissolved materials or of substances in suspension.

Table 7.1. *Typical chemical and mineral composition of common volcanic rock groups and commercial glass*

	Basalt (%)	Andesite (%)	Dacite (%)	Rhyolite (%)	Brown Bottle Glass (%)
SiO_2	51	54	64	74	68
Al_2O_3	14	17	17	13	2
Fe_3O_4	12	9	5	2	2
MgO	6	4	2	0.3	2
CaO	10	8	6	1	9
Na_2O	2	4	4	3	14
K_2O	0.8	1	2	5	0.4
Major minerals	Olivine Pyroxene Ca-Feldspar	Pyroxene Amphibole Ca, Na-Feldspar	Amphibole Biotite Quartz Na-Feldspar	Biotite Quartz K-Feldspar Na-Feldspar	None (all glass)

When magma or molten lava is cooled quickly – in seconds or minutes as compared to days or years – there is not enough time for new, minute crystals of silicate minerals to form, and the resulting rock is all or partly glass. Rapidly cooled lava will form solid glass; shiny black obsidian is a good example. In contrast, lava ponded into thick flows will cool slowly and its molten glass will crystallize into small mineral grains. Because natural glass is high in iron compared to manufactured glass it is generally opaque. Manufactured glass is also high in soda compared to lava glass; soda lowers the melting temperature and apparently retards the cooling glass from crystallizing into tiny silicate minerals.

Volcanic gases bubble out of molten lava under conditions of low pressure at the Earth's surface. Additional gas is released from molten glass as it begins to crystallize. Some volcanic rock is so filled with tiny gas-bubble holes that it has a special name – pumice. In this case the composition name becomes an adjective: for example, basaltic pumice or dacitic pumice.

Mineral content is the real key to volcanic rock types. Armed with a small magnifying lens and a knowledge of mineralogy, a field geologist can make quite good identifications. However, confirmation by chemical analysis and mineral identification under a polarizing microscope are essential.

Table 7.1 shows the key mineral and chemical compositions of the four common volcanic rock types considered in this chapter. The rock types – basalt through rhyolite – apply to all forms of volcanic products from lava flows to ashfalls and pyroclastic flows. These more explosive products, covered in the next two chapters, are more likely to form from silica-rich magmas.

8

Volcanic fallout

Kodiak, Alaska: June 6, 1912

The brilliant sunshine of early June was almost enough to make the residents of remote Kodiak Island forget the cold and dark of the Alaskan winter just past. Except for a few earthquakes, which are not unusual in volcano country, the only sign the islanders had of any ominous happenings were the "sun dogs," or halos around the sun. As the day wore on the brightness of the sky seemed to dim, and even though June was the month of the midnight sun, the sky darkened alarmingly.

As people opened doors to look at the sky, the smell of sulfur and the sight of gritty ash falling out of the heavy air made it clear that even though they had heard no sound, there had been a major volcanic eruption somewhere on the Alaskan peninsula.

Longtime Kodiak resident W. F. Erskine in 1962 compiled an account of the calamity from the letters and journals left to him by his parents. As one fisherman wrote to his wife,

My Dear Wife Tania:

. . . A mountain has burst near here . . . We are covered with ashes, in some places, 10 feet and 6 feet deep. All this began on the 6th of June. Night and day we light lanterns. We cannot see the daylight. . . . We are expecting death at any moment, and we have no water. All the streams are covered with ashes. Just ashes mixed with water. Here are darkness and hell. . . . And noise. I do not know whether it is day or night. . . . So kissing and blessing you both, goodbye. Forgive me. Perhaps we shall see each other again. God is merciful. Pray for us.

Your husband,   Ivan Orloff

The earth is trembling; it lightens every minute. It is terrible. We are praying.

The reference to lightens probably means lightning bolts. Static electricity on ash particles in an eruption cloud is discharged in large and numerous lightning strikes.

The ship's log of the mail steamer *Dora* en route to Kodiak contains this entry:

(June 6) At 6 P.M. passed through Uzinka Narrows, fine and clear ahead, and continued on expecting to make Kodiak. At 6:30 P.M., when off Spruce Rock, which is about 3 ½ miles from Mill Bay Rocks and the entrance to Kodiak, ashes commenced to fall and in a few minutes we were in complete darkness, not even the water over the ship's side could be seen.

I continued on in hopes that I might pick up entrance to Kodiak, but when vessel had run the distance by the log, conditions were the same, so I decided to head out to sea and get clear of all danger. At 7:22 I set a course NE. by N. (magnetic). Wind commenced to increase rapidly now from the southwest and vessel was driven before it. Heavy thunder and lightning commenced early in the afternoon and continued through the night. Birds of all species kept falling on the deck in a helpless condition. The temperature rose owing to the heat of the volcanic ash, the latter permeating into all parts of the ship, even down into the engine-room.

About 4:30 A.M. the next day vessel cleared the black smoke, emerging into a fiery red haze, which turned yellow, and by 6:00 A.M. the ashes had ceased to fall and the horizon was perfectly clear from west to north. . . .

During the day Katmai continued to be emitting columes of smoke and could be seen at a distance of over 100 miles.

The vessel was covered with ash from trucks to deck, the decks having ashes from 4 to 6 inches deep.

Two major ashfalls smothered Kodiak on June 6 and 7 (Fig. 8.1). People could not see a lantern held at arm's length in what should have been daytime, but everyone groped their way to some shelter. Miraculously no one was killed, and when the air cleared on June 8, everyone in Kodiak crowded onto steamers and fishing boats and left town. As no further ashfalls occurred, some ships returned. Nellie Erskine described the sight in a letter to her mother:

The next day, the tenth, I went over to Kodiak to take a look. Poor old Kodiak, it certainly is a wreck. Whether the people can live here is not at all settled. Of course it will take time and patience. It certainly

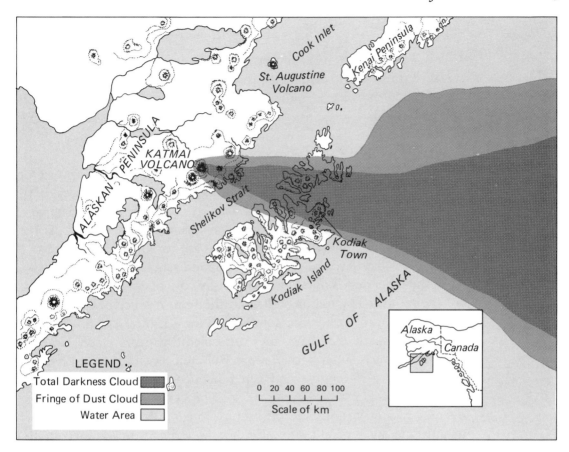

Figure 8.1. Path of the ash cloud from the 1912 eruption of Katmai Volcano. The town of Kodiak was covered by about 60 cm of fine volcanic ash, which in places was piled into drifts as much as 2 meters thick. (After W. F. Erskine, Katmai *[New York: Abelard Schuman, 1962], p. 10)*

is awfully discouraging, but we are not worrying. The feeling of thankfulness that we were saved is too strong yet. The ashes are about two feet on the level but in places it is higher than your head. People are dazed and dirty. They are despondent still. But I guess we can make something out of it if we try real hard.

Luckily for the people of Kodiak, the island's main industry was fishing, not agriculture, so its recovery was fairly rapid. The drifts of yellowish gray ash were eventually incorporated into the soil, and wildlife gradually returned. Although there had been fears that the pumice-filled water would change the spawning patterns of the salmon and halibut, the fishing was virtually unharmed.

ASH, BLOCKS, AND BOMBS

Volcanic ash is both bane and boon. Thick deposits of falling ash have smothered towns and farmlands, forcing whole populations to move or starve, yet in the long term the slow decomposition of that nutrient-rich volcanic fallout is responsible for some of the world's best soils. The irony is that the fertility of fields surrounding active volcanoes lures people to live in the shadow of destruction.

Although the name "volcanic ash" implies that it is a product of combustion by fire, it is instead the fine fragments of volcanic rock, both glass and minerals, that are blasted from explosive eruptions. The very tiny particles, less than $\frac{1}{16}$ millimeter in diameter, are called volcanic dust. Technically, the particles designated ash vary from about $\frac{1}{16}$ to 2 millimeters, and have the texture of common sand.

Of the other debris hurled into the air by volcanoes, cinders, or scoria, are the more coarse fragments (again the term has been borrowed incorrectly from the residue of combustion fire); if these small lumps are still molten when they land, they are called "spatter." Pieces larger – sometimes much larger – than a baseball are called blocks and bombs. Blocks are generally angular and were solid when ejected; volcanic bombs are plastic in flight. Blocks are usually older rocks broken by the explosive opening of a new vent; volcanic bombs, in contrast, are chunks of new magma and are generally still incandescent and soft during their flight. Some bombs acquire peculiar twisted shapes as they spin through the air; others, called "bread-crust bombs," have a cracked and separated crust that has cooled and hardened in flight and has been pulled apart by the still-expanding gases in the hot plastic interior; those that are soft enough to flatten on landing are called "cow-dung bombs" by certain bucolic geologists.

Materials composed of fragments of volcanic rock are known as "pyroclastic deposits," and may originate in one of two ways: by being hurled into the air and falling back to Earth, or by rapid lateral movement of pyroclastic material along the ground surface in the form of an extremely destructive *nuée ardente*. Chapter 9 deals with these pyroclastic flows in detail; this chapter concentrates on the volcanic debris that falls from the sky, called "ashfall" or "block-fall" deposits.

The most important process in forming various types of volcanic fallout is the way in which the particles moving through the air are naturally sorted. Just as farmers once winnowed grain from chaff by tossing it up into a light breeze, the force of gravity and the resistance of air sort volcanic debris by its size and density. Cinders land close to their source, whereas if the wind is blowing, light volcanic ash may land many

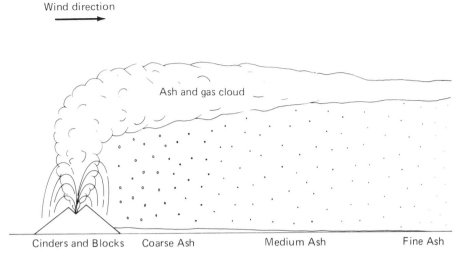

Wind direction

Ash and gas cloud

Cinders and Blocks   Coarse Ash   Medium Ash   Fine Ash

Figure 8.2. Cinders hurled from a volcanic vent fall to form a cone surrounding the vent. Coarse ash particles fall out downwind near the vent as progressively finer ash falls in a downwind band at greater distances.

hundreds of kilometers downwind from its source (Fig. 8.2). Volcanic dust and aerosols (tiny droplets of volcanic gas and water) are sometimes injected into the stratosphere by a large eruption, and may stay suspended in the atmosphere to circle the globe for months or years. Blocks and bombs sometimes defy this general rule that coarse fragments land close by the volcano and finer materials are carried to greater distances. If the blocks and bombs are ejected at an angle from the crater – like cannon shells from a howitzer – their high speed and weight overcome air resistance, and they may land several kilometers from the vent.

Ashfall deposits often blanket the ground like snowfall, covering hills and valleys alike with a layer of material (Figs. 8.3, 8.4). The combination of good sorting and the blanketlike nature of falling fragments provide the main evidence to distinguish between ashfall deposits and pyroclastic-flow deposits. This distinction is important to a volcanologist in trying to reconstruct the eruptive habits of a volcano by studying its prehistoric deposits.

ARENAL VOLCANO

Specific volcanic eruptions provide good examples of the various kinds of fallout material from blocks to dust. Beginning with large fragments, an awesome barrage of huge volcanic blocks and bombs were produced by the 1968 eruption of a small stratovolcano in Costa Rica called Arenal,

Figure 8.3. Ash from the 1943–52 eruption of Parícutin Volcano in Mexico blanketed the surrounding land so deeply that the trees were killed. The gulleys in the ash layers were cut by runoff during heavy rains. (Photograph by K. Segerstrom, U.S. Geological Survey)

killing many people and leaving the surrounding countryside looking like a battlefield. Arenal Volcano had no record of eruption before its sudden explosions in July 1968. An earthquake swarm occurred for ten hours before the eruption, but no other indications warned of the forthcoming calamity.

The first explosions produced pyroclastic flows and immense showers of large and small blocks and bombs, devastating an area of about 12 square kilometers. Many of the bombs were incandescent and disintegrated when they hit the ground, leaving large craters as much as 50 meters wide. In the area between two small towns, Tabacon and Pueblo Nuevo, the devastation was complete, with bomb craters everywhere. Some trees were still standing, but their leaves and smaller branches had been stripped away. Four thousand people evacuated the area in panic; many were injured and 78 were killed. A few weaker explosions of ash occurred during the next two months, and then thick block lava flows of andesite began to issue from the lowest explosion vent. Eruptive

Figure 8.4. The inclination of these ash layers from Oshima Volcano in Japan reflects the sloping ground surface on which they were deposited. Here they are exposed in a road cut.

activity, mainly lava flows and a few small explosions, has continued at Arenal since the initial spectacular outbreak.

William Melson of the Smithsonian Institution investigated the bombs and blocks that were hurled out in the early explosions at Arenal. Many were thrown as far as 5 kilometers from their vent and were obviously large, but because they broke into hundreds of smaller fragments on impact the original sizes are unknown. To reach the distances these bombs and blocks were hurled, their initial velocities would have had to be as much as 600 meters per second – more than 2,000 kilometers per

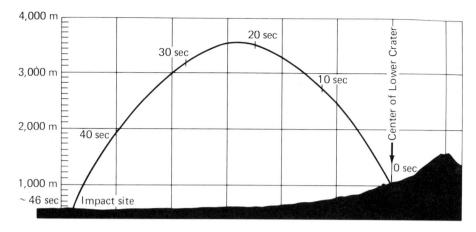

Figure 8.5. Large blocks thrown out by volcanic explosions follow a parabolic path. At Arenal Volcano in Costa Rica the 1968 eruption showered the surrounding area with large blocks like cannon shells. In this example a block landing more than 5 km from the vent was airborne for 46 seconds. (From William Melson, Smithsonian Institution)

hour. The trajectories of their flight indicate they were propelled upward at steep angles like mortar shells, reaching heights of 3 kilometers above their vent and traveling for 40 to 50 seconds before crashing to the ground (Fig. 8.5). On landing, the larger volcanic bombs and blocks dug huge impact craters, tens of meters wide and 3 to 4 meters deep (Fig. 8.6). Many smaller craters, a meter or two across with the projectile block still intact inside, also dotted the devastated area.

CERRO NEGRO VOLCANO

Sometimes an eruption will produce several types of volcanic rocks and loose deposits at the same time; Nicaragua's Cerro Negro Volcano's frequent eruptions in the 1960s and 1970s are good examples. Cerro Negro's cinder cone, a few hundred meters high and still growing, was born in 1850 and has been intermittently erupting basaltic products since that time (see Fig. 1.6). During 1968, Cerro Negro erupted cinders, ash, blocks, and a lava flow all at the same time.

Closely spaced small explosions from Cerro Negro's central crater hurled out dark clouds of ash and cinders. The cinders piled up on the cone around the vent, while the finer ash blew downwind to form a gray blanket across the tropical green hills. At night the spray of incandescent fragments that were building the cinder cone could be seen within the dark explosion clouds. However, by day the darker ash cloud obscured the lava fountains from the central vent. Sometimes larger bursts would

Figure 8.6. Blocks hurled out by the 1968 explosive eruption of Arenal Volcano in Costa Rica formed large impact craters. (Photograph by La Nacíon, San Jose, Costa Rica)

throw out blocks of lava that would land high on the cinder cone, then roll and skip down the steep slope of loose cinders to pile up near its base. While this pyroclastic action was coming from the central crater, a small vent at the base of the cinder cone was erupting an aa lava flow.

The complex activity at Cerro Negro illustrates the difficulty of trying to classify volcanic eruptions into specific types. It does, however, show how eruption processes can produce very different volcanic materials, depending on the gas content of the erupting magma. In this case, higher gas content in the central vent broke up the erupting lava into fragments that were then sorted by wind and gravity into separate deposits of ash, cinders, and blocks. After losing some of its gas content from the central vent, molten lava leaked through a crack in the cinder cone to issue as a flow from a vent at the base of the cone.

The ashfalls from Cerro Negro are usually a few centimeters thick and blow downwind over narrow sectors of the countryside. Although they were devastating to some individual farmers, they had little impact on the overall economy of the region. Unfortunately, this is not always true with large ashfalls from giant volcanic eruptions, which can force the migration of entire populations. Pyroclastic flows from the 1815 eruption of Tambora Volcano in Indonesia killed more than 10,000 people. The thick ashfall buried the rice fields and 80,000 more died in the ensuing famine. Although the region near Katmai Volcano in Alaska is nearly unpopulated, the desolation caused by the thick deposits of volcanic ash and pyroclastic flows in 1912 is still dramatically evident.

KATMAI VOLCANO

The eruption at Katmai lasted only 60 hours, but in that time it disgorged more than 30 cubic kilometers of rhyolitic to andesitic rock fragments, with two-thirds of this volume forming ashfalls and the remaining third forming a thick pyroclastic flow of the type described in Chapter 9. Mapping the ashfalls has shown them to be eight or nine separate layers, two of which are very thick and extend far downwind. Two calderas were formed by this huge eruption; one near Novarupta, which was the source of most of the magma, and the other engulfing the old summit of Mount Katmai to create a cliff-walled basin 3 kilometers wide and 1 kilometer deep.

The fallout deposits are thick blankets of white to gray pumice lumps and finer ash, blown mainly eastward by moderately strong winds (Fig. 8.7). The size of the ash particles decreases and the deposits become thinner with increasing distance from the erupting vents. Figures 8.1 and 8.8 show the distribution of ash from this enormous eruption. Much of it landed in the ocean, and traces of Katmai ash have been found in ice cores from Greenland. The total calculated volume of the Katmai ashfalls is about 20 cubic kilometers. This was formed from about 9 cubic kilometers of magma, with the difference in volume accounted for by the low density of the expanded pumice and the pore space between fragments. If all the ashfall from this huge eruption of Katmai were piled into a pyramid, it would be nearly 3 kilometers high, 20 times taller than the largest pyramid in Egypt.

Figure 8.7. Ash from the great Katmai eruption in 1912 still blankets the land. The fuming peak in this aerial view is Trident Volcano, which erupted in the 1950s and 1960s, covering the buff-colored 1912 ash layers with dark andesitic lava.

A volcanic ashfall when first deposited covers a landscape fairly evenly with a blanket of particles; before long, however, landslides, high winds, and heavy rains will transport much of this loose material into thicker deposits in valley bottoms and at the mouths of streams. Many streams and rivers east of Katmai are still so overloaded with volcanic ash that they have formed deadly quicksand deposits, and are nearly impossible to cross.

VOLCANIC DUST ENCIRCLES THE EARTH

The dust from large volcanic explosions – especially from Plinian eruptions that can jet clouds of debris to high altitudes for many hours – stays suspended in the stratosphere for months to years. Along with this dust, which is composed of tiny glass and mineral particles, some major eruptions vent large amounts of water vapor and sulfur gases into the

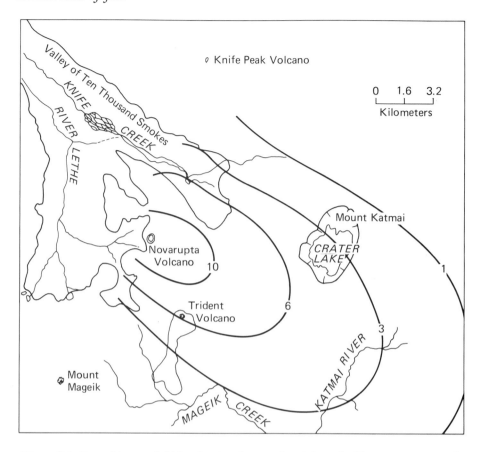

Figure 8.8. Several layers of thick volcanic ash were ejected from the Novarupta vent on the west flank of Mount Katmai during its brief eruption in 1912. Layer C shown here is more than 10 meters thick near its source and diminishes to 1 meter in thickness 20 km southeast of Novarupta. (After G. H. Curtis, Studies in Volcanology, Geological Society of America Memoir 116: 177)

upper atmosphere. These gases combine to form an aerosol of very small droplets of sulphuric acid. Above the clouds in the stratosphere – which generally starts about 15 kilometers above sea level – the dust and aerosol particles are not washed down by rainfall and thus can stay suspended for long periods of time. High-speed winds in the stratosphere spread the injections of volcanic dust and aerosols into layers that can encircle the Earth; colorful sunsets and halos around the Moon and Sun are manifestations of those fine particles and aerosols high in the atmosphere.

Another important effect of such high haze-layers is their ability to absorb energy from incoming sunlight. This causes the stratosphere to warm up and the Earth's surface to cool. How much this may cause changes in weather and climate is still a matter of debate among scientists, but there is circumstantial evidence that periods of cooler temperature worldwide have followed some major volcanic eruptions for a year or two. The eruption of El Chichón Volcano in Mexico in 1982 is an interesting example of a well-studied eruption injecting major amounts of dust and aerosol into the stratosphere. Chapter 12 discusses in more detail the eruption of El Chichón and its possible effect on weather and climate.

Volcanic fallout deposits ranging from fine dust to large blocks, combined with the pyroclastic flows considered in Chapter 9, form the bulk of volcanic eruption deposits worldwide. Although the common perception is that lava flows are the principle volcanic products, in volcanoes related to subduction zones the ratio of fragmental deposits to lava flows generally exceeds 90 percent. Volcanoes on rift zones, such as those in Iceland and East Africa, average about 40 percent fragmental materials; and those in Hawaii, only about 1 percent; but because subduction-related volcanoes are the most numerous type, about 80 percent of volcanic products on land are either fallout debris or pyroclastic flows. If deep submarine volcanoes, whose products are thought to be lava flows, were included, the ratio may be closer to 50 percent.

It is evident from the discussion thus far that volcanology is not a narrow field of study. Volcanoes have their roots deep beneath the Earth's surface, and some of their products are lofted into the stratosphere. They are a natural laboratory where scientists from many fields – geologists, chemists, physicists, botanists, and meteorologists – are needed to study their complex processes and effects.

9

Pyroclastic flows, avalanches, and mudflows

Bay of Naples, Italy: August 24, A.D. 79

Vesuvius roared to life like an angry giant awakened from a thousand-year sleep, catching unaware most residents of Pompeii and Hercula-neum. Some probably had time to escape, but most were undecided whether to go or stay when huge glowing avalanches swept down the volcano's flanks, engulfing their cities and burying them as if in a time capsule. Sealed and forgotten, it was not until 1709 that the ancient cities were accidentally rediscovered by a well-digger, and one of the world's richest archaeological – and geological – investigations began. Nearly 2,000 years after the volcanic burial, excavations are continuing and important discoveries are being made.

The most vivid account of Vesuvius's disaster – and probably the earliest recorded description of a volcanic eruption – was written at the time by Pliny the Younger. In his letter to Tacitus he tells of the eruption and the death of his uncle, the great Roman naturalist Pliny the Elder, who was also commander of the Roman fleet and had decided to go to the site to investigate the phenomenon.

> [At] about one in the afternoon, my mother desired him to observe a cloud which had appeared of a very unusual size and shape...he immediately arose from his books and went out upon a rising ground from whence he might get a better sight of this very uncommon appearance. A cloud, from which mountain was uncertain at this distance, was ascending, the form of which I cannot give you a more exact description of than by likening it to that of a pine tree, for it shot up in the form of a very tall trunk, which spread itself out at the top into several branches; occasioned, I imagine, either by a sudden gust of air that impelled it, the force of which decreased as it advanced upwards,

Plate 1. A fountain of molten lava dwarfs a man in a heat-resistant suit. This eruption of Krafla Volcano in Iceland is similar to those of Hawaiian Volcanoes. (Photograph by Katia Krafft)

Plate 2. (Overleaf) A river of molten lava, with a temperature of 1,100°C, pours from the Puu Oo vent on the east rift of Kilauea Volcano in Hawaii. (Photograph by Katia Krafft)

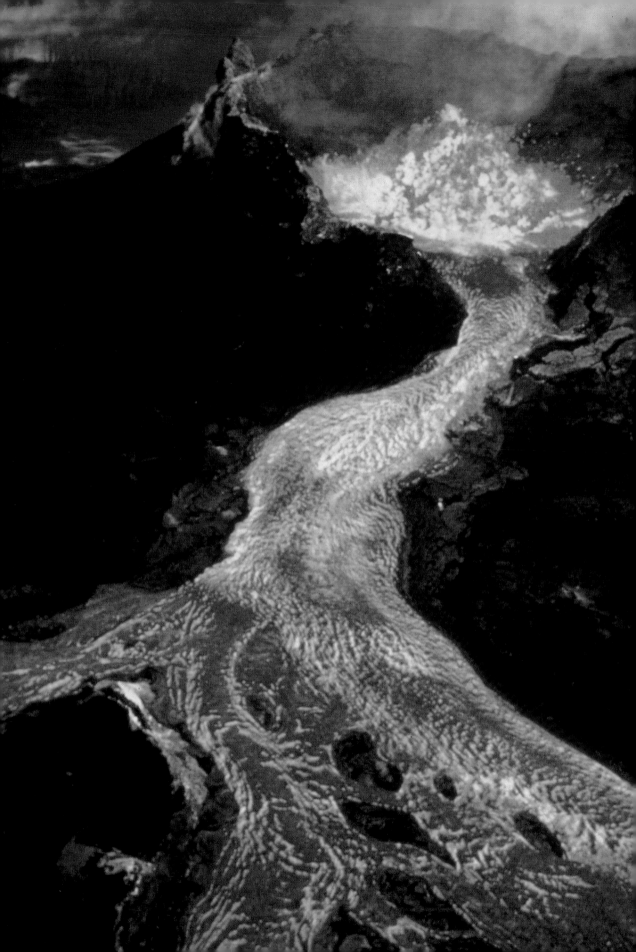

Plates 3 and 4. Lava flows in Hawaii are of two major types. A pahoehoe flow (above) is generally thin and has a smooth or ropy surface. An aa flow (below) is thicker and has a top layer of rough broken fragments. Both flows are basaltic lava of the same chemical composition. (Top photograph by D. A. Swanson, U.S. Geological Survey; bottom photograph by J. D. Griggs, U.S. Geological Survey)

Plate 5. Lava spatter falls around the erupting vent, building a 200-meter-high cinder and spatter cone called Puu Oo (Hill of the Oo Bird) on the east rift of Kilauea Volcano in Hawaii. (Photograph by George Ulrich, U.S. Geological Survey)

Plate 6. Waves move across the surface of an active lava lake in Halemaumau Crater during a summit eruption of Kilauea Volcano in Hawaii. (Photograph by Russ Apple, U.S. National Park Service)

Plates 7 and 8. *Temperature of molten lava can be measured directly by a thermocouple pushed into the flow (above) or by an optical pyrometer that determines the temperature by the color of the flow as seen through a hole in the roof of an active lava tube (below). The temperature of erupting lavas in Hawaii ranges from about 1,100° to 1,200°C. (Top photograph by Robert Christiansen, U.S. Geological Survey; bottom photograph by Robin Holcomb, U.S. Geological Survey)*

Plate 9. Molten lava in Hawaii surges through lava tubes beneath the surface of pahoehoe flows for several kilometers before spilling into the sea. (Photograph by D. W. Peterson, U.S. Geological Survey)

Plate 10. At the start of many Hawaiian eruptions, molten lava spurts from long fissures that break their way to the surface and form "curtains of fire." Geologist sampling the newly erupted lava deflects incandescent blobs with a metal shield that is generally used to reflect the intense radiated heat. (Photograph by J. D. Griggs, U.S. Geological Survey)

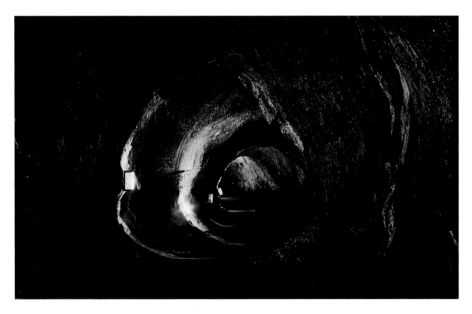

Plate 11. When a Hawaiian eruption stops or changes location, molten lava drains from tunnels beneath the surface of the flows, forming long, sinuous caves a few meters in diameter and thousands of meters long called "lava tubes." The Thurston Lava Tube in Hawaii Volcanoes National Park is a popular place to walk through a now-cooled lava cave. (Photograph by Jane Takahashi, U.S. Geological Survey)

Plate 12. During the Mauna Ulu (Growing Mountain) eruption in Hawaii, lava fountains sprayed hundreds of meters into the air. The flows from these fountains crossed nearly level ground and then cascaded into nearby craters formed during earlier activity. (Photograph by D. A. Swanson, U.S. Geological Survey)

Plate 13. Gas bubbles from the red-hot molten rock at the shore of a Hawaiian lava lake. (Photograph by D. A. Swanson, U.S. Geological Survey)

Plate 15. *Mount St. Helens as it appeared in June 1970. This 2,950-meter-high stratovolcano had been dormant since 1857.*

Plate 16. *During the avalanche and explosive eruption of May 18, 1980, Mount St. Helens's summit lost more than 400 meters in elevation and a great horseshoe-shaped crater 2 km wide, 3 km long, and 600 meters deep was blasted out of the remaining mountain stump. A dome of viscous lava more than 250 meters high and a kilometer in diameter has squeezed up into the crater during many small eruptions since 1980. (Photograph by Lyn Topinka, U.S. Geological Survey)*

Plate 14. *(Overleaf) The giant eruption cloud from Mount St. Helens on May 18, 1980, was too big to photograph. An impressive, but much smaller, explosion cloud from Mount St. Helens seen in this photograph taken July 22, 1980, reached a height of 13 km. (Photograph by Katia Krafft)*

Plate 17. *Katmai Volcano in Alaska erupted 30 cubic km of ash and pyroclastic flows in 1912, filling the 20-km-long, 3-km-wide valley in this photograph with hot volcanic fragments. Over the years the Valley of Ten Thousand Smokes has cooled to a barren plain cut by new stream canyons. The fume cloud on the horizon is from Mount Trident, a nearby volcano active during the 1960s. (Photograph by U.S. National Park Service)*

Plate 18. *Volcanic ash deposits exposed in a road cut on Oshima Island in Japan are evidence of past major explosive eruptions. Erosion removed much of the lower ash layers before the upper layers blanketed the area.*

Plate 19. Steam and hot water gush from Old Faithful Geyser in Yellowstone National Park. Although its timing is not accurate enough to set your watch by, Old Faithful sprays 30 to 50 meters high about every hour. No volcanic eruptions have occurred at Yellowstone in recorded history, but the park's geysers and hot springs remind visitors that its fires are not extinct.

Plate 20. The summit of Ruapehu Volcano in New Zealand contains a large lake of hot water. The lake was thrown out of the crater during an eruption in 1953, and the resulting mud flows swept away a railroad bridge, wrecking an express train and causing many deaths.

Plate 21. Algae flourish on the surface of a warm lake near Tarawera Volcano in New Zealand. Many thermal features in this area were formed by a 16-km-long fissure eruption in 1886.

Plate 22. Mud pots at Rotorua, New Zealand, are formed by steam leaking up from underground geothermal reservoirs. The volcanic belt of the North Island is the site of several huge explosive eruptions during prehistoric but "recent" geologic time.

Plate 23. (Opposite) Sulfur deposits accumulate around many volcanic gas vents, but the choking smell of brimstone keeps most of the curious away. These deposits in Indonesia are mined for their sulfur content. (Photograph by Katia Krafft)

Plate 24. (Below) Undersea hot springs called "black smokers" pour out jets of water at temperatures exceeding 300°C. They are found along the mid-ocean ridges at depths of 2,500 meters, where the great pressure of the ocean prevents formation of steam. Iron and zinc sulfide particles, precipitated as the hot water is cooled by the surrounding ocean water, form the black smoke. (Scripps Institution of Oceanography photograph, provided by R. A. Koski, U.S. Geological Survey)

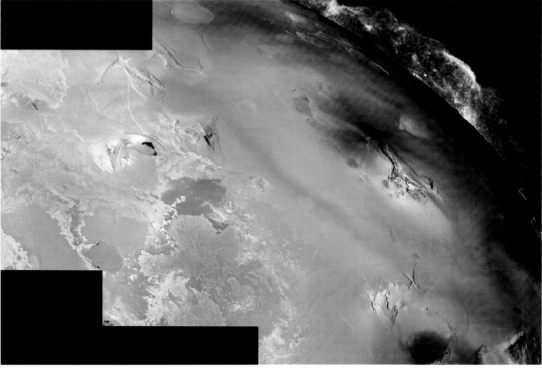

Plates 25 and 26. Io, one of the moons of Jupiter, appears to have many violently active volcanoes. Molten sulfur and sulfur gases are thought to be major products of these strange volcanic eruptions. Major portions of the 3,500-km-diameter moon (above), *one-quarter the size of Earth, are covered by yellow to red deposits. A closer look* (below) *shows a volcano named Pele* (right center) *spouting a veil of particles 300 km high into the near-vacuum atmosphere of Io. (Images by NASA and Alfred McEwen of the U.S. Geological Survey)*

Plate 27. (Overleaf) *An explosion of Anak Krakatau in 1979 ejects a spray of incandescent lava fragments. (Photograph by Katia Krafft)*

or the cloud itself being pressed back again by its own weight . . . it appeared sometimes bright and sometimes dark and spotted, according as it was either more or less impregnated with earth and cinders. . . .

He ordered the galleys to put to sea, and went himself on board with an intention of assisting . . . the several towns which lay thickly strewn along the beautiful coast. . . .

He was now so close to the mountain that the cinders, which grew thicker and hotter the nearer he approached, fell into the ships, together with pumice stones. . . . Here he stopped to consider whether he should turn back again; to which the pilot advising him "Fortune" he said, "favors the brave. . . ."

The wind was favorable for carrying my uncle to Pomponianus at Stabiae, whom he found in the greatest consternation; he embraced him tenderly, encouraging and urging him to keep up his spirits, and the more effectually to sooth his fears by seeming unconcerned himself, ordered a bath to be got ready, and then, after having bathed, sat down to supper with great cheerfulness, or at least (which is just as heroic) with every appearance of it.

Meanwhile broad flames shone out in several places from Mount Vesusius, which the darkness of night contributed to render still brighter and clearer. . . . They consulted together whether it would be most prudent to trust to the houses, which now rocked from side to side with frequent and violent concussions as though shaken from their very foundations; or fly to the open fields where the calcined stones and cinders, though light indeed, yet fell in large showers and threatened destruction. In this choice of dangers they resolved for the fields. . . . They went out then, having pillows tied upon their heads with napkins; and this was their whole defense against the storm of stones that fell round them.

It was now day everywhere else, but there a deeper darkness prevailed than in the thickest night . . . my uncle, laying himself down upon a sail cloth . . . called twice for some cold water, which he drank, when immediately the flames, preceded by a strong whiff of sulphur, dispersed the rest of the party, and obliged him to rise. He raised himself up with the assistance of two of his servants, and instantly fell down dead; suffocated, as I conjecture, by some gross and noxious vapor. . . . As soon as it was light again, which was not till the third day after this melancholy accident, his body was found entire, and without any marks of violence upon it, in the dress in which he fell, and looking more like a man asleep than dead. . . . During all this time my mother and I, who were at Misenum – but this has no connection with your history . . . so I will end here. . . . Farewell.*

* Bosanquet, F. C. T., ed., *Pliny's Letters* (London: George Bell and Sons, 1903), 194–8.

Figure 9.1. The ruins of Italy's Pompeii have been excavated from their burial beneath 2 to 3 meters of pumice fall that was followed by another 2 meters of pyroclastic flow deposits. The A.D. 79 eruption of Mount Vesuvius, seen in the background 10 km away, killed about 2,000 people in Pompeii.

The eruption of Vesuvius in A.D. 79 produced major pumice-and-ash falls, followed by pyroclastic flows. Unknown thousands of people were killed. Pompeii was blanketed by thick pumice-fall deposits, which in turn were covered by pyroclastic flows, the whole section totaling 4 to 5 meters thick (Figs. 9.1, 9.2). Herculaneum was upwind from the early falling debris but was inundated by multiple pyroclastic flows, which total 10 to 20 meters in thickness. Stabiae (now Castellammare) was blanketed by more than a meter of fallout of ash and pumice lumps, but only the edge of the last major pyroclastic flow reached there. It was that hot blast (and perhaps an overweight man's weak heart) that killed Pliny the Elder. His servants lived to tell the story to Pliny the Younger.

The most awesome and destructive volcanic eruptions are those that generate pyroclastic flows, avalanches, and mudflows. Any one of these fast-moving phenomena can be catastrophic and each is quite different, both in manner of emplacement and in resulting deposits.

Figure 9.2. Most people killed at Pompeii had survived the pumice fall but were overwhelmed several hours later by pyroclastic flows. Their bodies left holes in deposits, which, when filled with plaster during the excavations, give a dramatic picture of the terror that occurred there more than 1,900 years ago. (Photograph by Katia Krafft)

PYROCLASTIC FLOWS

When in motion pyroclastic flows are mixtures of hot volcanic fragments and swirling gases. These mixtures are more dense than air and flow along the ground; they are mobilized by the turbulent gases that keep the solid fragments in suspension. There is no familiar analogy for a fluidized flow, but the net result is that this turbulent mixture of gas and suspended solid particles behaves like a liquid with less viscosity than water.

The energy that drives a pyroclastic flow comes from expanding gases and from gravity. The gases may be in the rapidly expanding volcanic explosion cloud, and in this case the pyroclastic flow may be pushed in some initial direction by the force of a lateral blast. In addition, expanding gases issue from the newly erupted, hot-but-solid volcanic fragments, and can also form by the heating of air mixed into the flow as it moves forward.

Following the initial push and internal expansion, the pull of gravity is the main factor that causes these heavier-than-air flows to sweep down-slope. Many pyroclastic flows form by the fallback of volcanic ash clouds that originally jetted upward, but were so dense that gravity overcame the initial upward thrust causing them to collapse.

Not only are the processes that mobilize these glowing avalanches complex, but their deposits are diverse. Much of the fragmental material is the size of ash, and many geologists refer to pyroclastic flows as ash flows. However, in this type of flow lumps and blocks of pumice and more solid rocks from the vent area are often swept along and incorporated, making their final deposits look like concrete – large fragments surrounded by fine-grained material. Sometimes pyroclastic deposits are so thick and hot when they come to rest that the pumice lumps are squashed into flat, glassy lenses, and the whole deposit is welded into solid rock. These welded tuffs often form during great caldera-forming eruptions.

Because of the complexity of both the process of formation and the diversity of deposits, there are many names for types of pyroclastic flows: *nuée ardente,* glowing avalanche, base surge, surge, ash flow, and pumice flow. The loose deposits or overall rock unit formed by pyroclastic flow processes are sometimes called *nuée ardente* deposits. Tuff (a soft material), welded tuff (hard), and ignimbrite are additional terms for the deposits of a pyroclastic flow.

Part of the diversity of nomenclature arises because the differences between types of eruptions and volcanic products are complex and apt to be gradational rather than distinct. Nevertheless, some generalities are possible. Pyroclastic flow deposits tend to pond into low topographic areas instead of forming a blanket over hills, slopes, and valleys as do volcanic ashfall deposits; also, pyroclastic flow deposits usually are poorly sorted with regard to fragment size and generally are not layered.

Two types of pyroclastic flows were produced during the 1980 eruption of Mount St. Helens (described in Chapter 2) (Fig. 9.3). The initial blast formed a ground-hugging, dark cloud of old shattered rock, hot fragments from the exploding shallow magma intrusion and its surrounding hydrothermal system, and splintered trees that were swept down by the advancing blast cloud (Fig. 9.4). On hilltops and gentle slopes this deposit is generally less than one meter thick. Where it was plastered on steep slopes it was still mobile enough to slide down and fill the valley bottoms to much greater thicknesses. Later in the day, after the high Plinian eruption cloud had formed, pyroclastic flows spilled out

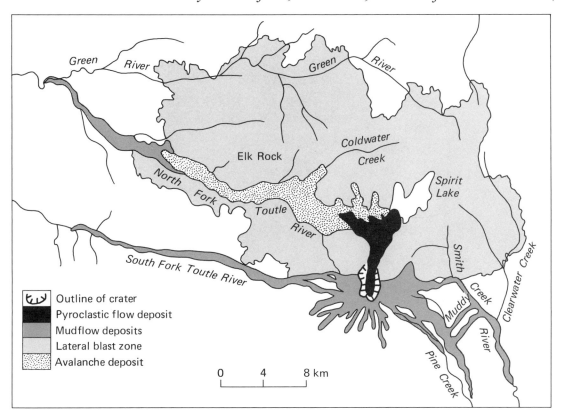

Figure 9.3. Map showing the major deposits and devastated areas from the May 18, 1980, eruption of Mount St. Helens. Area covered by ashfall is not included. (From Robert Tilling, Eruptions of Mount St. Helens: Past Present and Future, *U.S. Geological Survey, 1984)*

of the main vent. Reinforced by fallback from the eruption cloud, they swept down the breached north face of Mount St. Helens. This mixture of hot ash and pumice blocks piled into thick deposits at the base of the mountain. Where these deposits ran into Spirit Lake or covered wet areas, secondary steam explosions added to the overall chaos.

Pyroclastic flows have temperatures that range from about 100° to 800° C. Most are dry, but if they cool to 100° C while still moving, steam begins to condense to hot water and wet particles may become plastered on objects in their path. A large, hot pyroclastic flow overwhelms almost everything in its path. The mass and high temperature combined with speeds as high as 200 kilometers per hour are a nearly irresistible force. (Figure 3.2 illustrates what a flow of this type did to the city of Saint Pierre during a few fateful minutes.)

About one-third of the 15 cubic kilometers of rhyolitic to dacitic magma that was erupted at Mount Katmai in 1912 produced a large ash flow that swept 20 kilometers down a wide glacial valley, filling it with

Figure 9.4. Millions of trees were blown down by the lateral blast during the first few minutes of the May 18, 1980, eruption of Mount St. Helens. (Photograph by U.S. Geological Survey)

hot ash and pumice to depths of as much as 200 meters. Water beneath the pyroclastic flow deposit was heated to steam, and that seeped upward through the ash and pumice. This steam, and other volcanic gases leaking from the deposit, escaped to the surface through thousands of fumaroles on the flat barren surface of the newly filled valley. When explorers first reached this still-fuming, remote place in 1916, four years after the eruption, they named it the Valley of Ten Thousand Smokes. By now most of the deposit has cooled; only a few fumaroles still remain.

Figure 9.5. Valley of Ten Thousand Smokes as seen in 1916 when the thick pyroclastic flow was still hot. View is from Katmai Pass. (Photograph by Robert Griggs, National Geographic Society)

The explosive eruption at Katmai was the largest in this century, but its volume – 15 cubic kilometers of magma expanded to 30 cubic kilometers of fallout debris and ash-flow deposits – is small compared to some prehistoric eruptions (Figs. 9.5 and 9.6). Three gigantic eruptions at Yellowstone in the western United States have built a high plateau of thick pyroclastic deposits. One of these huge caldera-forming eruptions occurred 2 million years ago and erupted 3,000 cubic kilometers of rhyolitic magma, an eruption 200 times larger than Katmai.

AVALANCHES

Large avalanches from the flanks of volcanoes were not well studied before the 1980 eruption of Mount St. Helens. As described in Chapter

Figure 9.6. Pyroclastic flow formed in the 1912 eruption of Katmai Volcano, Alaska, cut through by stream erosion. This 25-meter-thick section is 20 km down-valley from the vent. Notice the irregular bottom of the flow where it covered mounds of glacial deposits, in contrast to the nearly flat layers within the flow. (Photograph by Peter Ward)

2, between late March and mid-May of that year a shallow intrusion of magma beneath the cone of Mount St. Helens tilted the north face of the mountain upward and outward. This oversteepened the slope so much that it broke loose on May 18 and formed a huge landslide of rock, ice, snow, soil, and trees called a "debris avalanche." One branch of the slide swept into Spirit Lake, causing waves more than 200 meters high; another overtopped a 400-meter-high ridge north of the base of the main cone of Mount St. Helens, while the bulk of the slide raced 23 kilometers down the valley of the North Fork of the Toutle River at speeds of 100 to 200 kilometers per hour. The avalanche, which contained both ice and hot rocks, had an estimated average temperature near 100° C. Expanding steam may have helped mobilize the avalanche.

The resulting deposit is a chaotic mixture of rocks and loose debris, in places as thick as 200 meters but averaging about 45 meters (Fig.

Figure 9.7. The 3-cubic-km avalanche from Mount St. Helens roared 23 km down the Toutle River Valley in just a few minutes. Notice the hummocky surface of the deposit, characteristic of large avalanches. (Photograph by Lyn Topinka, U.S. Geological Survey, 1985)

9.7). The surface of the debris avalanche is covered by hundreds of hummocks and hollows with a relief of about 20 meters. The volume of the avalanche is nearly 3 cubic kilometers, and accounts for most of the material lost when the huge horseshoe-shaped crater that replaced the peak of Mount St. Helens was blasted out. The rest of the excavation of the crater was accomplished in several ways: by the great explosion caused when the avalanche slid away, suddenly releasing pressure on the shallow magma and its surrounding hydrothermal system; by the jetting of ash; and later by ash flows from the exposed magma conduit.

The avalanche at Mount St. Helens was among the largest ever recorded, but in the last few years even larger prehistoric avalanches associated with volcanoes have been recognized. The widespread area of jumbled hills north of Mount Shasta in California is now interpreted to be the deposit of an immense volcanic avalanche that roared down Shasta about 300,000 years ago. It had a volume of more than 25 cubic kilometers, and traveled 43 kilometers northwest of the mountain's base.

MUDFLOWS

Mudflows are mobilized by floods of water and, like avalanches, are common in, but not restricted to, volcanic areas (Fig. 9.8). They can travel long distances from their source, often endangering lives and destroying property far beyond the reach of pyroclastic flows and avalanches.

Mudflows were another major component of the Mount St. Helens eruption. They consisted of a slurry of ash and fine rock particles mixed with water, and had the consistency of wet cement. The ash blanket from the eruption and the crushed rock in the avalanche deposit provided the solid matter; the water came from a combination of melting snow and ice, the spillover from Spirit Lake, and the compaction of the wet avalanche deposit squeezing water out.

The largest mudflow came down the North Fork of the Toutle River, below the avalanche, and crested more than ten hours after the eruption began. High mud marks indicated that it exceeded historic flood levels by 9 meters (Fig. 9.9). Hundreds of large logs stored at a logging camp were swept into the mudflow and battered downstream bridges into twisted wreckage. The flood of mud also destroyed many homes near the river. Downstream from the Toutle River, mudflow deposits spilling into the Columbia River clogged the navigation channel and marooned oceangoing ships for many weeks until the passage could be dredged clear again.

When surging down steep valleys, a mudflow can travel at speeds as high as 30 to 40 kilometers per hour and can carry huge boulders in its dense currents. Old deposits exposed in road cuts and riverbanks often reveal a mixture of rounded cobbles in a matrix of fine particles. Although most volcanic mudflows occur during an eruption, there have been cases

Figure 9.8. Small eruption of the lava dome in Bezymianny Volcano, Kamchatka, rapidly melts the snow and generates mudflows. The great explosive eruption of Bezymianny in 1955 followed by dome growth is similar to the recent activity of Mount St. Helens. (Photograph from the Institute of Volcanology, Kamchatka, USSR)

where they have been generated by torrential rains pouring down on loose volcanic ash deposits from previous eruptions.

Even when a volcano is not in eruption, the alteration of volcanic rocks to clay by hydrothermal action and fumaroles can take place. If a large volume of these soft and slippery clay deposits form high on the flanks of a steep volcanic cone, they can be dislodged by heavy rains or an earthquake to form avalanches, mudflows, or both.

The 1985 destruction of Armero, the Colombian town of 25,000

Figure 9.9. The massive mudflow that swept down the Toutle River after the eruption of Mount St. Helens reached much higher than any previously recorded flood of that river. Mud marks on these trees are nearly 10 meters above ground level. (U.S. Geological Survey photograph)

people described in Chapter 1, is a tragic example of a mudflow disaster that could have been prevented by advance planning and present technology. It took nearly two hours for the mudflow formed by sudden melting of the ice field high on Ruiz Volcano to reach Armero. Simple flood and mudflow warning devices, such as those used extensively in Japan, which consist of trip wires suspended across stream valleys and connected by radio to downstream police stations, might have saved more than 20,000 lives at Armero.

VOLCANIC DEPOSITS AND ASSESSING VOLCANIC HAZARDS

Volcanic rocks and deposits of volcanic debris are nature's record of volcanic eruptions. By mapping such formations to determine their character, extent, and age, a geologist can roughly reconstruct the past eruptive habits of a volcano. It helps immensely to study the rocks and deposits from well-documented historic eruptions, because in these cases a cause-and-effect relationship between the process and the resulting formations can be clearly established. In prehistoric eruptions, only the rocks and deposits are available, and sometimes they have been eroded away or covered by younger deposits. Nevertheless, careful mapping and knowledge of volcanic processes and products yields reasonable interpretations of a volcano's geologic history. One of the maxims of geology is that the present is the key to the past. In terms of future volcanic hazards, it can also be said that the past is the key to the future. Part III of this book explores the impact of volcanoes on human history.

PART III

Volcanic risk and reward

Civilization exists by geological consent, subject to change without
notice.

– Will Durant

10

Volcanic catastrophes

Lower Nyos, Cameroon: August 21, 1986

Most of the villagers of Lower Nyos were sleeping and didn't hear the rumbling explosion at Lake Nyos, a mile up the valley. Those who were awake, perhaps finishing a late supper after a busy market day, heard it but had no way of knowing that the noise signaled the release from the lake of a huge cloud of deadly gas. The poisonous cloud poured silently down the valley, snuffing out the lives of 1,700 people, asleep or awake. In the village of Lower Nyos alone more than 1,200 people died, but 5 or 6 – inexplicably – survived. They told of family members eating and talking one moment and dropping dead the next. One woman woke in the morning to find her five children dead around her.

The deadly cloud, 50 meters thick, moved quietly down the valley for a distance 16 kilometers, killing another 500 people in neighboring villages before it dissipated. The tragedy was compounded by the fact that August 21 had been a market day and herders from the hills had come to town with their cattle to barter with lowland farmers. Most were camping overnight near town and perished along with their cattle; in the morning the meadows near Nyos were littered with 3,000 dead animals (Fig.10.1).

Rescue workers who reached the town a few days later said it looked as if a neutron bomb had hit; houses and gardens were undamaged, but dead bodies were everywhere. No birds sang and no flies swarmed around the dead – the obliteration of life had been complete.

The source of the lethal fumes that caused this incredible catastrophe was Lake Nyos, a small but deep lake that fills a volcanic crater. Over hundreds of years toxic gases – chiefly carbon dioxide – leaked up from the volcano and accumulated in deep layers of the lake water, which became supersaturated with dissolved gas. This gas accumulation prob-

Figure 10.1. Thousands of cattle were suffocated by the carbon dioxide eruption from Lake Nyos in Cameroon, Africa, August 21, 1986. (Photograph by Jack Lockwood, U.S. Geological Survey)

ably occurs at volcanic lakes all over the world, but most are in temperate zones where the water is stirred and mixed by seasonal temperature variations or storm winds. In Cameroon the surface lake-water, constantly warmed by the tropical sun and thus less dense, stays at the top instead of mixing. The colder water at the bottom of the lake stays below, absorbing more and more gas over the years.

Something happened that August night to trigger the release of the huge gas cloud from the lake, but no one is certain what it was. A chilly rain had fallen all that day, and perhaps enough cold rainwater had poured into the lake to cause it to overturn. The heavy rains could also have triggered a landslide that stirred the lake, or an earthquake may have disturbed the delicate equilibrium. Some researchers even suggest that the gas was blown from beneath the lake bed by a small volcanic explosion, although no evidence has been found to confirm that theory.

Whatever the immediate cause, the results were devastating. The deadly cloud, heavier than air, hugged the ground as it traveled down the valley from the lake and suffocated virtually every living thing in its path until it was dispersed by winds and rain. This catastrophe not only killed nearly 2,000 people, but it displaced thousands more who were terrified of living near the "killer lake" but had no place else to go.

Catastrophe is a subjective word, now used to describe troubles ranging from personal to cosmic. In a geological sense it refers more specifically to a sudden, violent change in the physical condition of the Earth's surface, and one that affects its inhabitants; it is in that way the word is used here. In this chapter *volcanic catastrophe* means an eruption devastating enough to change the established social order of an entire region, or to change the way in which volcanic activity is understood.

The eruption in about 1600 B.C. of Thera, an island in the Aegean Sea near Greece, apparently contributed to the decline of Minoan power on Crete and the rise of Mycenaean power on mainland Greece, thus affecting the entire early course of Western civilization. A major eruption near the present city of San Salvador in Central America in about A.D. 300 changed the direction of Mayan civilization, while the eruptions of Tambora and Krakatau in Indonesia in the nineteenth century killed thousands of people and produced worldwide atmospheric effects that have led scientists to realize that an individual volcanic eruption can affect the entire globe.

THERA

The island of Thera, sometimes called Santorini, lies in the Aegean Sea 120 kilometers north of Crete (Fig. 10.2). This 15-kilometer-wide volcanic island, now part of Greece, apparently played a major role both in the mythology and in the early evolution of Western civilization. Before about 1600 B.C., Thera was an important part of the Minoan culture. Centered on Crete, the Minoans were a powerful seafaring nation that traded goods throughout the Eastern Mediterranean.

The Minoans established a major city, now known as Akroteri, on

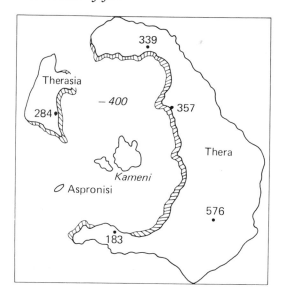

Figure 10.2. Map of Santorini Volcano, Greece. The huge explosive eruption and caldera collapse that occurred about 1600 B.C. may be the basis of the Atlantis legend. The Kameni Islands are lava domes and thick lava flows that have formed from many smaller eruptions between 197 B.C. and A.D. 1950. (Modified from Bullard, Volcanoes of the Earth)

Thera and by 1600 B.C. this prosperous seaport was a thriving commercial center. Then a major earthquake partially destroyed some of the houses. While the damage was still being cleaned up, some other event – another earthquake or perhaps ashfall from a small warning eruption – prompted the people of Akroteri to evacuate their city. Soon after the evacuation, the entire city was buried by an enormous volcanic eruption, and remained sealed until 1967, when archaeologists began investigating the area.

Excavations of Akroteri conducted by Greek archeologists have so far uncovered homes up to three stories high, containing beautiful ceramic jars and colorful frescoes painted on the walls (Fig. 10.3). The houses were built along paved streets with stone-lined sewers beneath the pavement. All excavations so far indicate that Akroteri was home to a wealthy and artistic people. The evacuation was not done in panic; precious metals, weapons, and tools were carried away and jars of seed grain were left behind to plant the fields again when people returned.

But they never came back. Sometime after they fled, presumably in ships, the volcanic core of Thera erupted with a gigantic blast and the collapse of a 7-kilometer-wide caldera, gutting the high center of the island. Akroteri was buried by more than 5 meters of pumice-fall and ash-flow deposits, and the center of the island collapsed to 300 meters below sea level.

Some 30 cubic kilometers of dacitic to rhyolitic magma was spewed out in huge ash clouds and pyroclastic flows. Ash layers from the eruption have been found in cores taken from the seafloor over wide areas of the

Figure 10.3. Akroteri, a prosperous Minoan city on the Aegean Island of Thera, was buried by pumice fall and pyroclastic flows during the explosive eruption about 1600 B.C.

Mediterranean southeast of Thera. When the collapsed caldera was filled by the inrushing sea, large tsunamis must have raced across the eastern Mediterranean and crashed into the shores of Crete, Greece, Turkey, and perhaps even Egypt and Syria.

Some scholars claim that Thera is the lost island of Atlantis, and that its engulfment in one day and night of earthquakes and floods is the real basis of the Atlantis legend that was related by Plato from Egyptian sources. Other students of the Bible and the history of ancient Egypt have suggested that the ten plagues of Egypt may have been caused by the ash fallout from Thera, and that the parting of the waters in the exodus of the Jews from Egypt may have been brought about by the ebb and flow of large tsunamis on the low coastal plain between Egypt and Sinai. It is only fair to say that not all historians associate the eruption of Thera with the Atlantis myth or the Book of Exodus, but the debate is lively. Whatever consensus, if any, that may evolve from this interesting controversy, will probably be more opinion than fact.

Today Thera is a semicircular island; a rocky rim surrounding a vast caldera. Volcanic domes that have grown during several eruptions over the past 2,000 years (the latest in 1950) form smaller islands that rise above the sea in the center of the cliff-rimmed caldera. Occasionally wisps of steam escape the central island, a reminder of the once violent scene of volcanic activity.

Minoan culture began to wane sometime after the Thera eruption, and the Mycenaean culture of mainland Greece began its ascendency. The golden age of Greece followed and provided much of the framework of Western civilization. It is interesting to speculate what language we might be writing and reading if Thera had not erupted.

Some scholars disagree that the eruption was this significant, pointing to wars and pestilence as more important causes of shifts in history. Nevertheless, a major natural disaster can alter the delicate balance of power between competing cultures or nations, and pestilence often follows. Of one thing there is no doubt: The explosive eruption and collapse of Thera that occurred about 1600 B.C. was a major natural disaster, one of the largest volcanic eruptions of the past 5,000 years.

ILOPANGO

San Salvador, the capital of El Salvador, is situated on the west edge of a large intermountain lake named Ilopango (Fig. 10.4). Much of the city is built on top of a thick deposit of volcanic ash called *tierra blanca* (white earth). Recent geologic studies have established that Lake Ilopango fills an 8- by 11-kilometer elliptical caldera that in about A.D. 300 erupted 15 cubic kilometers of dacitic magma in thick ashfall and ashflow deposits that form the *tierra blanca*.

Archaeologist Payson Sheets has concluded that this eruption was a catastrophe for the highland Mayan Indians living in El Salvador at that time. The resulting shift of population and trade routes, however, benefited the lowland Mayans living in the Petén and Yucatán areas, and stimulated the classic Mayan civilization that flourished in those areas from A.D. 300 to 900.

The eruption produced a combination of ashfall and pumice-fall layers, and thick ash flows that swept as far as 45 kilometers from the caldera. A half-meter of ashfall covered the Mayan city at Chalchuapa 75 kilometers northwest of Lake Ilopango, and deposits nearly 50 meters thick occur near the caldera rim.

Charred trees in the ash-flow deposits yield radiocarbon dates of A.D. 260, plus or minus a hundred years, and archaeological dating of artifacts beneath the ash and pumice layers also indicate a date of about A.D. 300. Like most explosions that result in caldera collapse, the eruption was

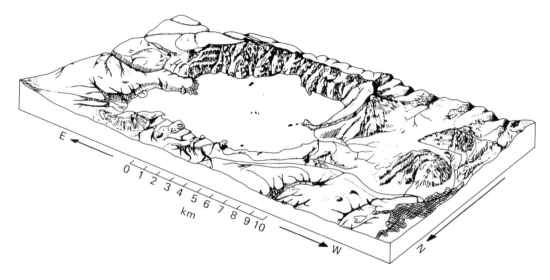

Figure 10.4. Diagram showing aerial view to the southwest of Lake Ilopango in El Salvador. This 8- by 11-km caldera disgorged 50 cubic km of ashfall and pyroclastic flows in a great eruption about A.D. 300. Migration of survivors from this region apparently stimulated the classic Mayan civilization in Guatemala and Yucatán in the centuries that followed. The present city of San Salvador is shown in the northwest corner of the diagram. The small islands in the center of the lake are part of a lava dome that grew there in 1880. (After H. Meyer-Abich, in Catalogue of Active Volcanoes of Central America, *International Volcanological Association, 1958)*

apparently extremely violent but short-lived, lasting only a few days. The initial ashfall was probably not fatal to most nearby inhabitants, and if they had fled before the devastating ash flows began, large numbers of refugees may have survived. Added to these would be thousands more who lived beyond the reach of the ash flows but whose fields were covered by more than 20 centimeters of the ashfall that blanketed most of El Salvador and possibly parts of Guatemala and Honduras as well.

No records are available, but estimates of persons killed and those permanently displaced range from thousands to hundreds of thousands. Many refugees moved into Tikal and other Mayan centers in the lowlands of northern Guatemala, Belize, and the Yucatán Peninsula of Mexico, 400 kilometers or more north of the Ilopango devastation. The Pacific coastal trade routes from Mexico were detoured through Tikal, and this influx of people and commerce seemed to spark the rise of the Mayan civilization in the lowlands of Central America.

El Salvador remained nearly abandoned for more than a hundred years, but farmers slowly began to return to work the thin new soil forming on the *tierra blanca* ashes. Unfortunately, the story did not end there. In about A.D. 600, ash from a smaller, more local eruption completely buried some of the area again. Archaeologists today are excavating

Figure 10.5. Volcanic ash layers below and above this buried farmhouse in El Salvador attest to repeated destruction and recovery from explosive eruptions. The Ilopango ash beneath the house spread over a wide area sometime during the third century A.D. *After about 200 years the ash had weathered to soil sufficiently that farmers began to return. In the sixth century a smaller but nearby eruption from Laguna Caldera destroyed and buried this Cerén site excavated in 1978. (Photograph by Payson D. Sheets, University of Colorado)*

hearths and walls of houses built on top of the Ilopango ash that were engulfed by another 4 meters of ash (Fig. 10.5).

The lives of individuals are fragile, but the human species is tenacious. People migrate out and back from areas of devastation in cycle after cycle of relentless destruction and hopeful renewal.

TAMBORA

About three years of small explosions preceded the great caldera-collapse eruption of Tambora in 1815. This stratovolcano, more than 4,000 meters high before the eruption, is on Sumbawa, the second island east of Bali in Indonesia (Fig. 10.6). Sumbawa was so devastated that most reports of the catastrophe were obtained from witnesses at sea or on other islands.

The climactic explosions, ashfalls, and ash flows occurred on April 10 and 11, with detonations heard as far as 1,500 kilometers distant in Sumatra and Ternate. Daytime darkness and clouds of falling ash reached East Java, 500 kilometers west of Tambora, and the Southern Celebes, 300 kilometers to the north. Small tsunamis, 1 to 2 meters high, reached neighboring islands, and thick blankets of "cinders" as much as 60 centimeters thick were seen floating on the sea.

The ash fallout on Bali, 250 kilometers west, was about 30 centimeters, and even thicker on the island of Lombok between Bali and Sumbawa. It smothered crops on Lombok and Sumbawa, and a great famine and plague added many fatalities to the death toll of the eruption. Estimates put the total fatalities at 60,000 to 90,000; some 10,000 killed by the eruption and the rest by starvation and disease.

Sumbawa is remote and not often visited by scientific expeditions, so only a few geologists have investigated the results of the eruption. They have discovered a 6-kilometer-wide jagged-edged caldera whose rim is about a kilometer lower than the pre-1815 summit. The floor of the deep caldera contains a lake about 700 meters below the rim. The pyroclastic deposits expelled in 1815 are unusually low in silica and high in potassium compared to most caldera-forming eruptions, and their volume has been variously estimated from 30 cubic kilometers to 150 cubic kilometers. The volume of magma needed to match the caldera volume is about 40 cubic kilometers, and until more measurements are made on the 1815 deposits on land and in sea cores, this figure seems to be a reasonable estimate. This makes Tambora one of the largest – perhaps the largest – eruptions in historic time.

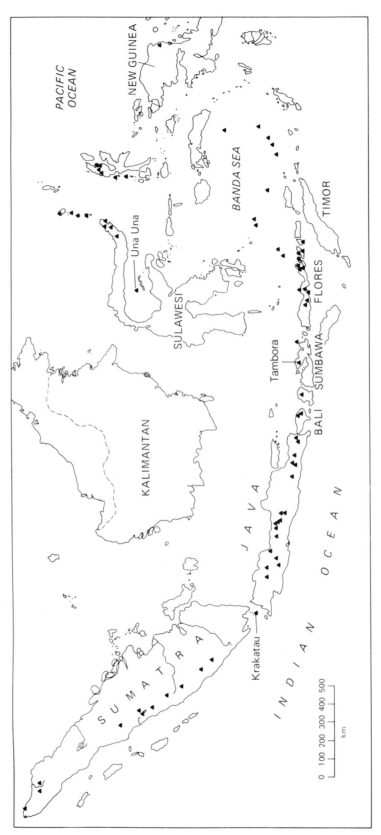

Figure 10.6. Map of Indonesia showing localities of 80 volcanoes that have erupted during historic time—a record high for any one country. The eruption of Krakatau in 1883 killed about 36,000 people; that of Tambora in 1815 caused an estimated 90,000 deaths. (Modified from Neumann van Padang, in Catalogue of Active Volcanoes of Indonesia, International Volcanological Association, 1951).

KRAKATAU

The 1883 eruption of Krakatau, sometimes called Krakatoa, an island volcano in Indonesia, is one of the world's most infamous natural catastrophes. After a few months of small, intermittent eruptions and a day of larger blasts, Krakatau unleashed nature's largest well-documented explosion. On August 27 the climactic detonation lofted ashes 50 kilometers high. The blast was heard in Australia more than 4,000 kilometers away, and produced a shock wave that registered on barographs around the world. Ashes fell over an area of 500,000 square kilometers, and the volcanic dust that was injected into the stratosphere encircled the Earth in two weeks.

Krakatau is an uninhabited group of islands in the Sunda Strait, between Java and Sumatra, and no one was killed by the direct effects of the explosion. However, as 6 cubic kilometers of dacitic magma spewed forth, an 8-kilometer-wide caldera collapsed to depths of more than 200 meters below sea level (Fig. 10.7). Tsunamis from this sudden displacement of the sea, some reaching heights exceeding 30 meters, swept over the low coastal plains of Java and Sumatra, killing 36,000 people.

It is often said that Krakatau "blew its top," but that is not strictly true. The erupted material was largely dacitic pumice, a rock so filled with bubble holes that it floats on water. Huge blankets of this floating pumice choked the Sunda Strait for weeks after the eruption. The older rock that formed much of this missing island was more solid andesite, and only a small percentage of these older rocks are found in the deposits from the 1883 eruption. Had the volcano literally blown its top, the deposits would be composed largely of broken fragments of this older rock. As in most caldera-forming eruptions, the basin was mainly formed by collapse into the evacuated magma chamber rather than by explosive excavation.

The global distribution of volcanic dust from Krakatau by high-speed, high-altitude winds gave meteorologists an early clue to the existence of jet streams they would not be able to measure directly for another 50 years. During the fall and winter of 1883, people the world over witnessed many strange and beautiful atmospheric effects. Sunsets were especially spectacular, inspiring Victorian artists and poets, including Tennyson, who wrote:

> Had the fierce ashes of some fiery peak
> Been hurled so high they ranged about the globe?
> For day by day, thro' many a blood-red eve,
> The wrathful sunset glared.

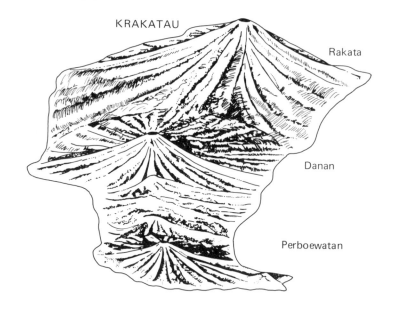

KRAKATAU

Rakata

Danan

Perboewatan

Figure 10.7. Before and after illustrations of the 1883 eruption of Krakatau in Indonesia. Danan and Perboewatan peaks were completely destroyed. The missing area was 23 square km. Where Danan once stood 450 meters high, the ocean is now 275 meters deep. Anak Krakatau (Child of Krakatau), seen in the lower picture, appeared above sea level in 1930 and has grown to be a new island about 2 km in diameter. The remnant of Rakata Island is seen to the left, behind Anak Krakatau. (Sketch from the Volcanological Survey of Indonesia)

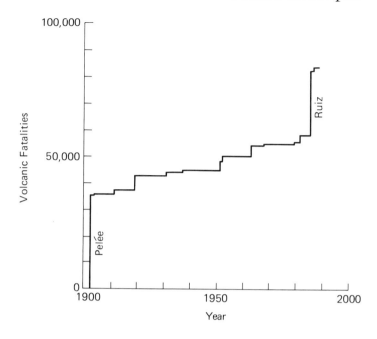

Figure 10.8. Cumulative number of deaths worldwide from volcanic eruptions since 1900. The two major jumps in the graph, in 1902 and 1985, result largely from the catastrophic eruptions of Mont Pelée in the Caribbean and Nevado del Ruiz in Colombia.

These worldwide effects of the great Krakatau eruption – both the shock waves measured by barographs and the dramatic atmospheric effects seen by millions – led scientists in Europe and America to recognize that the effects of a major volcanic eruption can be felt around the entire globe.

DEATHS AND DESTRUCTION

It has been estimated that more than 1 million people have been killed by volcanic eruptions in the past 2,000 years. It is not possible to document this number exactly because where destruction is worst, records and evidence are often destroyed as well. Better data exist for the past hundred years, and the toll appears to be about 100,000 deaths and $10 billion (U.S. dollar present value) in damage per century (Fig. 10.8). Most of the deaths occur in a few eruptions. In the twentieth century the 1902 *nuée ardente* eruption of Mont Pelée killed 29,000 people, and the mudflows from Ruiz Volcano in 1985 killed another 25,000. The two large eruptions in Indonesia in 1815 and 1883, Tambora and Krak-

Table 10.1. *Examples of large prehistoric explosive eruptions that formed calderas (only a small sample of many on Earth more than 10 km wide)*

Country	Caldera	Diameter (in kms)	Volume of erupted magma (in cubic km)	Time formed (in years ago)
Italy	Phlegraean Fields	13	50	35,000
New Zealand	Rotorua	15	200	140,000
Indonesia	Toba	100 × 35	2,800	75,000
Japan	Aira	20	110	22,000
United States	Yellowstone	60 × 45	1,000	630,000
United States	Long Valley	32 × 17	600	700,000
Guatemala	Atitlan	20 × 17	250	84,000

atau, with total fatalities of about 120,000, account for most of the deaths from volcanic action in the nineteenth century.

The historical record, however, probably does not provide a reasonable guide to future risk from volcanic hazards. There are two major reasons for this: First, the population of the Earth continues to increase rapidly, thereby putting many more people at risk from even moderate-scale eruptions. This was tragically illustrated by the mudflows from Ruiz Volcano in Colombia in 1985 that killed 22,000 people in Armero alone. More than a hundred years earlier, similar mudflows from Ruiz in the same area killed only about 1,000 people simply because fewer people lived in the hazardous zone.

Second, evidence shows that many volcanic eruptions in prehistoric times were much larger – 10 to 100 times greater in volume – than the huge eruptions of Thera and Tambora. Table 10.1 shows some examples of very large, explosive, caldera-forming eruptions that have occurred in various parts of the world in the last million years. The list in this table is by no means complete; current investigations in South America and elsewhere reveal evidence of other extremely large prehistoric eruptions. Based on present knowledge, it appears that eruptions of roughly 100 cubic kilometers occur on average about every 10,000 years, and those of 1,000 cubic kilometers, about every 100,000 years. If this is correct, infrequent but extremely large volcanic eruptions will occur in the future.

Places with the highest risk from volcanic eruptions are regions of potentially active explosive volcanoes that have high population densities. Italy, Indonesia, New Zealand, Papua New Guinea, the Philippines, Japan, the United States, Mexico, Central America, Colombia, Ecuador, Peru, and Chile are countries that should be concerned. The problem is not hopeless; some things that can be done to reduce volcanic risk are discussed in Chapter 11.

II

Forecasting volcanic eruptions

Mauna Loa, Hawaii: March 25, 1984

A red glow lit the night sky over Mauna Loa's summit caldera as that giant volcano awoke from a nine-year nap. At 1:30 A.M. the first lava fountains broke to the surface along a fissure, creating a "curtain of fire" across the caldera. As the crack extended and more fire fountains came into play, volcanic earthquakes and tremors were so strong that the astronomical telescopes on neighboring Mauna Kea Volcano, 50 kilometers away, could not be stabilized for observations.

During the first few hours, the eruption had migrated to vents at the edge of the caldera and started down the northeast rift zone. As vents at lower elevations opened, the ones that were higher upslope – including those in the caldera – gradually shut down. The main vents soon established themselves in a zone about a kilometer long, at the 2,900-meter level of the 4,169-meter mountain. The lava fountains rose to more than 20 meters, pouring out voluminous and fast-moving flows.

The main flow headed northeast toward Hilo, a port city 50 kilometers away. Moving swiftly and relentlessly, it covered 15 kilometers on the first day. Walls of cooling lava built up into levees on the edge of the flow, and the 1,130° C molten rock coursed down in a channel between them. As the flow advanced over the next few days its velocity decreased, but it was still confined to a narrow lobe aimed toward Hilo.

At night the red-hot flow and the fume clouds above it glowed orange from reflected light, and appeared from town to be even closer and more menacing than they were. The flow moved steadily down the slopes above Hilo, crushing and burning hundreds of acres of native rain forest. Explosions of methane gas that accumulated near the edges of the flow could be heard every few minutes, and sulfurous-smelling smoke hung in the air. Civil Defense officials and local residents began preparing for

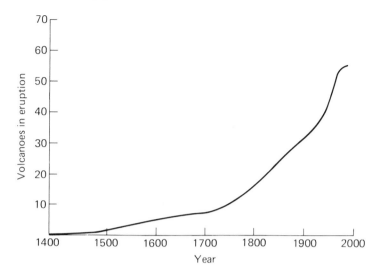

Figure 11.1. Number of volcanoes reported in eruption on a yearly basis since 1400. The curve is smoothed to a 30-year running average. The impression that the number of volcanic eruptions has been increasing is probably misleading. Growth of population, and improvements in global communication and record keeping, probably account for most of the increase. In the last decade the number of volcanoes, worldwide, reported to be in eruption has varied between 50 to 60 per year. (Data from Simkin et al., Volcanoes of the World, *1981; and Scientific Event Alert Network, Smithsonian Institution)*

the worst in case a shift in direction or speed of the flow made a sudden evacuation necessary.

By March 29 the advancing flow had reached to within 8 kilometers of the outskirts of town, but then a fortunate event occurred. About 15 kilometers upstream one of the levees confining the channel collapsed, and lava was diverted into a second, parallel flow. A few days later, just as the second flow caught up with the first, another levee break occurred. At the same time the lava production from the vents was slowing a bit, and the lava was becoming more viscous. In the days that followed more diversions took place even farther upslope, allowing the flows to spread out harmlessly over uninhabited land instead of advancing relentlessly downhill. By April 14 no active flows extended more than 2 kilometers from the vents. On April 15 the eruption ended, and the threat to Hilo was over.

Although the exact timing of the eruption came as something of a surprise, the eruption itself was not unexpected. Mauna Loa had last erupted in 1975, and before that in 1950. For the last ten years scientists at the Hawaiian Volcano Observatory had been carefully monitoring Mauna Loa's increasing earthquake activity, and measuring the inflation of the mountain as the magma chamber swelled beneath its summit. Both trends seemed to be accelerating in 1983, leading Observatory scientists to take the unusual step of publishing a forecast of a probable eruption within the next one to two years.

To be able to forecast or predict natural hazards such as earthquakes, volcanic eruptions, landslides, storms, floods, and climate change is one of the most important goals of the earth sciences. Weather forecasting is recognized as useful even though it is inexact, and volcanic eruption forecasting is used in that same sense here.

Reducing the risk of volcanic eruptions can be approached in three complementary but distinct ways: (1) by investigating the historic eruption record and prehistoric deposits of individual volcanoes; (2) by monitoring the vital signs of potentially active volcanoes; and (3) by educating the people living near volcanoes about what might occur in future eruptions and how they can reduce the potential dangers. An anecdote about eruption forecasting recommends a geologist to find out what *did* happen, a geophysicist to discover what *is* happening, and a politician to inform the public what *might* happen.

ERUPTIVE RECORD

Some volcanoes show quite consistent patterns of activity, while others are much more erratic. Naturally, when attempting to forecast future activity on the basis of a volcano's past habits, the more consistent performers are the most predictable. Nevertheless, conditions and patterns change, and all forecasts based on past performance are subject to some uncertainty. The historic record of volcanic eruptions is much better in some regions of the world than in others. Data has been kept in Italy and Greece for thousands of years, whereas in Hawaii the record is only about 200 years old.

Figure 11.1 is a graph of the number of known volcanic eruptions, worldwide, over the past six centuries. At first glance this plot suggests an alarming increase in volcanic eruptions; however, a more reasonable interpretation is that as global communications have improved, so has the reporting of volcanic eruptions.

Geologists can extend the record of volcanic activity back into prehistoric time by studying the character, sequence, and age of volcanic products erupted from individual volcanoes. This method is similar to the way in which archaeological excavation can establish the prehistoric record of human activities. By mapping the type and extent of lava flows, ash layers, mudflows, or other volcanic deposits, and by establishing their ages by radiocarbon or other dating techniques, geologists can decipher the prehistoric record of major eruptions over many thousands, some-

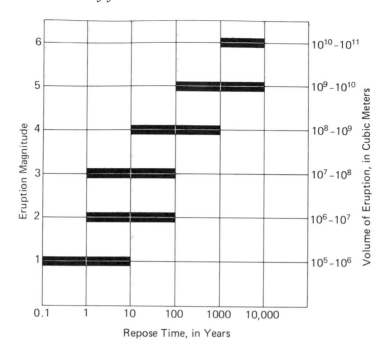

Figure 11.2. Repose times – the intervals between eruptions – are generally much longer before explosive eruptions of large magnitude. The bars in this graph represent the span of repose times of the majority of historical eruptions worldwide of that scale. Magnitude numbers indicate tenfold increases in the volume of pyroclastics ejected during an eruption; for example, an explosive eruption of magnitude 5 disgorges 1 cubic km, whereas a magnitude 6 disgorges 10 cubic km. Larger eruptions happen less frequently than smaller ones. Only eight magnitude 6 to 7 eruptions have happened in historic time, while nearly seven hundred magnitude 3 to 4 eruptions have occurred. (Data from Smithsonian Institution)

times millions, of years. The problem with this approach is that a small eruption may not produce deposits that can be easily mapped. Also, all or part of the record can be wiped out by major erosional periods like the ice ages. Careful geologic mapping can indicate only the larger patterns of a volcano's past activity. Nevertheless, this is extremely valuable; recorded history is so nearsighted to the long vistas of geologic time that it may see only the sand grains, not the beach.

Most of the 1,300 or more potentially active volcanoes on Earth have not been mapped to assess their prehistoric eruption record. Fewer than 10 percent have been investigated with regard to future volcanic hazards, and only about 30 volcanoes have comprehensive hazard maps and reports.

One problem in assessing the hazards from future eruptions is that many extremely dangerous volcanoes have lain dormant through historic time; some of the most destructive eruptions in the historic record have occurred at volcanoes that had been dormant for hundreds to thousands of years (Fig. 11.2). Many of these potentially violent volcanoes are considered extinct by the people living nearby, and because they are relatively unknown they are poorly studied in comparison to more notorious volcanoes that erupt frequently. Even volcanoes with well-documented records of many historical eruptions may show wide variations in repose times – a volcano's interval of quiet between periods of activity – and in the character of those eruptions. For example, Asama Volcano in Japan has erupted thousands of times since records of its activity began in the sixth century A.D. Since 1900, Asama's shortest repose times have been less than one day and the longest, more than five years. Most recent eruptions have expelled moderate explosions of ash, but a disasterous eruption in 1783 produced large ash-and-pumice falls, pyroclastic flows, and mudflows.

Despite these problems, some forecasts based on historic records and geologic mapping have been remarkably accurate, as was shown at Mount St. Helens. In 1978, Dwight Crandell and Donal Mullineaux of the U.S. Geological Survey published a hazards assessment in which they concluded that during the preceding 4,500 years this stratovolcano had been more active and more explosive than any other in the contiguous 48 states. During that time Mount St. Helens produced lava domes and flows, ash-and-pumice falls, pyroclastic flows, and mudflows, with an average repose period of 225 years. On the basis of their study of this past behavior, their report stated that Mount St. Helens would erupt again, "perhaps before the end of this century." Only two years later that forecast became reality (see Chapter 2).

Analysis of the duration of repose periods at a volcano with many recorded eruptions may reveal patterns that can aid in forecasting the onset of future eruptions. For example, the average repose time of Kilauea Volcano is between one and two years, with a pattern of groups of eruptions clustering together in time. Because of this pattern, the expected repose time before a future eruption is shorter than average during a cluster and longer than average if there have been no eruptions for several years. The lengthening repose time indicates that Kilauea has probably moved away from a period of clustered eruptions. Hekla Volcano in Iceland has just the opposite pattern; its average repose period is about 50 years, with the liklihood of an eruption increasing with each passing year. At Hekla the buildup of magma for the next eruption apparently continues throughout its repose period.

MONITORING

In order for a volcano to erupt, magma must move to the surface. These underground movements or a change in volume of molten rock generally produce signals that can be detected by geological, geophysical, and geochemical observations before an eruption begins. In some cases the changes may be obvious to people living near a volcano; for example, they may feel a swarm of earthquakes or notice consistently increased fuming from the crater. More often, the changes in earthquake pattern, ground surface deformation, and the composition of gas emissions are small and subtle and can be detected only by continuous monitoring with sensitive instruments. This is the job of a volcano observatory.

Establishing and maintaining a volcano observatory is expensive compared to determining the eruptive record and hazards of a volcano by geologic mapping, but these approaches to forecasting eruptions are complementary. The few volcanoes that have been thoroughly mapped and are being constantly monitored have the best record of successful forecasts (Fig. 11.3).

Monitoring active volcanoes involves continuous or frequent observation of earthquakes, ground surface deformation, the temperature and composition of fumaroles or crater lakes, and visual changes. During an eruption it is important to monitor the composition of products and the mass and rate of their emission. Other techniques, including electrical, magnetic, and gravity measurements, may also reveal changes in the subsurface structure or dynamics of the volcano under observation. Changes in any of the data – for example, the patterns of earthquakes, the rate of surface uplift, or the ratio of chlorine to sulfur in the gas from fumaroles – reveal information that is vital to understanding how the volcano works and to forecasting what may happen next.

An increase in earthquakes at shallow depths beneath a volcano is generally an indication of some significant change, perhaps an increase in volume of magma in the storage chambers (Fig. 11.4). Some earthquake swarms may be caused by changes in the regional stress system and are concentrated beneath a potentially active volcano because the hot rocks there are weaker and more prone to break than are colder rocks in the surrounding region. In fact, some magma chambers beneath volcanoes are so plastic that they deform without the fracturing that causes earthquakes. In this case, locating the envelope of tiny earthquake centers in the weak but still brittle rocks that surround a magma chamber provides a subsurface image of the location and size of that reservoir (Fig. 11.5).

Besides recording individual earthquakes, seismometers near an active volcano often pick up continuous ground vibrations called volcanic tremor. This tremor is nearly always seen during volcanic eruptions, and it is often recorded before an eruption starts. The causes of volcanic tremor probably include magma moving through conduits at depth, gases

Figure 11.3. The Hawaiian Volcano Observatory and Jaggar Museum on the rim of Kilauea Caldera. Halemaumau Crater is in the background in this aerial view to the south-southeast. (Photograph by J. D. Griggs, U.S. Geological Survey)

boiling out of magma or groundwater, or a close sequence of many small earthquakes. Because some volcanic tremor is caused by shallow intrusion of magma, it is of great importance in the seismic monitoring of active volcanoes.

Deformation of the ground surface also provides clues to subsurface volcanic activity. Using the Hawaiian volcanoes as examples again, magma moves upward nearly continuously and accumulates in the shallow magma reservoirs under Kilauea and Mauna Loa volcanoes, causing

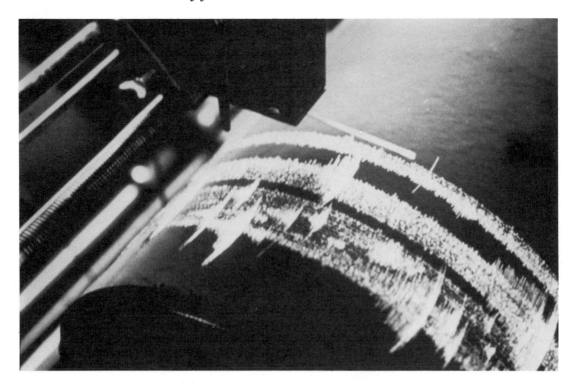

Figure 11.4. *Seismograph recording volcanic tremor (the continuous wiggly line) and earth-quake swarm (umbrella-shaped signals) occurring beneath Kilauea Volcano in Hawaii. The drum rotates at 1 mm per second, so some of the larger earthquake signals (about magnitude 3) die out slowly over a period of two to three minutes.*

uplift and stretching of their summit regions (Fig. 11.6). This defor-mation is spread over large areas and is detectable only by sensitive tiltmeters or surveying measurements. The pattern of the deformation depends on the depth to the swelling magma chamber; the deeper the source, the more widespread and smaller in amount the surface defor-mation. In Hawaii the figures for these depths agree with the earthquake evidence for shallow magma storage reservoirs beneath the calderas of Kilauea and Mauna Loa at depths of about 3 to 4 kilometers.

In addition to this subsurface information, the amount of surface swelling is an important factor in estimating the onset of a new eruption.

Figure 11.5. A: *Aerial view of Kilauea Caledera, Hawaii, looking north. The inner crater at left center is Halemaumau. Mauna Kea Volcano is in the distance. B: The locations of earthquakes (black dots) beneath the caldera region recorded from 1970 to 1983 envelop the inferred magma reservoir – the dashed oval. (Photograph from U.S. Geological Survey; earth-quake data from Fred Klein, U.S. Geological Survey)*

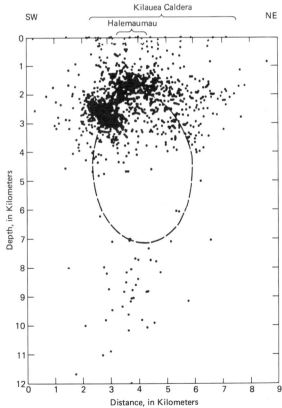

SW

Kilauea Caldera

Halemaumau

NE

Depth, in Kilometers

Distance, in Kilometers

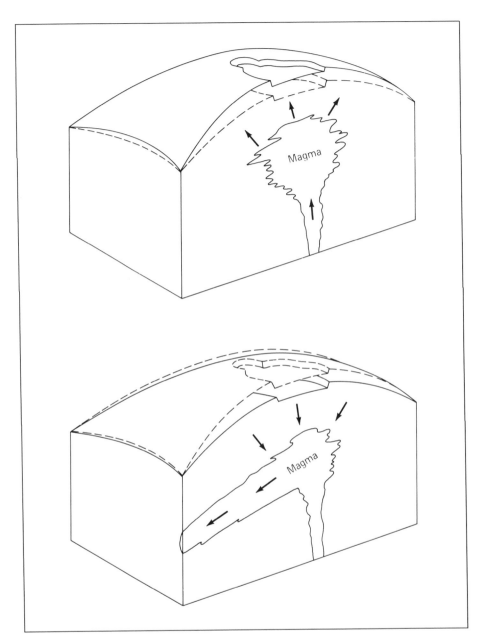

Figure 11.6. Diagrams showing the slow inflation (outward tilt) and rapid deflation (inward tilt) of the summit area of a Hawaiian-type volcano as magma accumulates and then escapes from the summit magma reservoir. The crack filled with moving magma (lower diagram) often reaches the surface at a lower elevation on the volcano and forms a flank eruption. The actual amount of uplift and subsidence of the apex of the deformation generally varies from a few centimeters to a few meters. (After Robert Tilling, U.S. Geological Survey, and Jean-Louis Cheminee, Institut de Physique du Globe de Paris)

High levels and high rates of surface uplift significantly increase the probability of eruption of both Kilauea and Mauna Loa.

Gases dissolved in magma begin to bubble out at different depths depending on their amounts and overall composition. For example, carbon dioxide (CO_2) is often the first to exsolve, with sulfur dioxide (SO_2) and hydrogen chloride (HCl) gases boiling out at the lower pressures of more shallow depths. An increase in the carbon-to-sulfur ratio at a fumarole may indicate that a new batch of magma has risen to shallow depths and CO_2 is preferentially leaking from this magma to the surface. Other factors such as changes in groundwater chemistry may also cause gas ratios to change; thus, although the interpretation of gas emission data by itself is not a certain indication of volcanic activity, it is most useful in reinforcing other clues.

Increases in the temperature of fumaroles and crater lakes have been observed before eruptions of volcanoes in Iceland and the Philippines, although again that is not conclusive evidence because decreases in rainfall can also increase the temperature of groundwater in volcanic areas. No single monitoring technique can clearly establish what is going on beneath a volcano or forecast its next move. The situation is similar to medical diagnosis: the more tests performed, the clearer the cause and prognosis, but the certainty is seldom 100 percent. However, a few key tests interpreted with good, experienced judgment are often better than an entire battery of tests without careful interpretation.

Monitoring techniques developed at one volcano observatory, though tailored to the specific volcano being studied, are useful at other locations as well. For example, the earthquake, surface deformation, and gas analysis techniques developed by volcanologists in Hawaii were immediately useful during the crisis at Mount St. Helens Volcano, showing that a shallow body of magma was being intruded beneath the north flank of the mountain. The date and magnitude of the huge May 18 eruption were not predicted, but the intense and continuous earthquake swarm, along with the high rate at which the bulge was growing on the north face before the major eruption formed the basis of the recommendation that access to the volcano be restricted. Sadly, some lives were lost, but the toll could have been much worse if no monitoring had taken place.

An ideal forecast of volcanic activity should include the location, timing, character, and magnitude of the potential eruption, and an accurate estimate of the uncertainty of each of these factors. The timing forecast should include not only the beginning of the eruption but also the time of maximum activity and when the eruption should stop. This is a large order, and the present state of the art is far from this goal. Nevertheless, forecasts based on probability can be greatly improved by more wide-spread use of existing knowledge, techniques, and instruments.

UNCERTAINTY AND PUBLIC POLICY

Exact predictions of the future activity of such complex structures as volcanoes may never be attained, because with every eruption the rules change. For example, at Kilauea Volcano the pressure level in the shallow magma chamber can be estimated quite accurately from ground defor-mation measurements, and thus the current pressure and the pressure level at which failure and eruption last occurred can be compared. How-ever, every time the volcano erupts its strength changes slightly, and this new strength can be determined only by the pressure level at which the next eruption occurs. Consequently, although there is a range of magma pressures at which an eruption is likely to start, no one specific pressure level can be counted on. As the level of inflation of Kilauea's summit increases, the probability of eruption increases also, but there is not a specific point at which an eruption can be expected to occur. In other words, Kilauea's changing strength can be determined only by hindsight, not by foresight.

Once the rocks around a slowly swelling magma chamber begin to fail and crack, earthquakes and volcanic tremor reveal this underground rupture and rapid injection of magma upward or outward from its hold-ing chamber. This type of warning may occur a few minutes to a few days before an eruption occurs. In some cases, however, the injection of magma into an underground crack does not reach the surface, and no eruption ensues. This is not a "false alarm" in the usual sense of the word; it would be more analogous to a fire alarm signaling a small fire that was under control before the engine arrived.

Officials responsible for areas that are at risk from volcanic activity may be ambivalent about wanting to know eruption forecasts that are not 100 percent certain. Imagine the dilemma of a mayor who has been told that his city is threatened by a 10 percent chance of a volcanic eruption of unknown magnitude sometime in the next month. Should an evacuation be ordered? Should the warning be ignored? A decision that involves a 10 percent chance of being a hero and a 90 percent chance of being a villain is an uncomfortable one to make. These social aspects of eruption forecasting are perhaps more difficult than the scientific investigations.

At the Ruiz Volcano disaster in Colombia in 1985, governing officials were warned that the town of Armero was built on top of a mudflow

Figure 11.7. A: *Photograph of hazard map showing predicted extent of mudflows near Armero in case of an eruption of Nevado del Ruiz Volcano in Colombia. This map was prepared and published before the November 13, 1985, eruption.* B: *Actual extent of mudflows that destroyed Armero during the November 13, 1985 eruption (see Chapter 1). (From Observatorio Vulcanologico de Colombia)*

Figure 11.8. Pyroclastic flows from Colo Volcano on Una Una Island, Sulawesi, Indonesia, destroyed the extensive coconut plantations that were the basis of the island's economy. However, the 7,000 residents of Una Una had been evacuated by boat just before the July and August 1983 explosions, and no one was killed. (Photograph by Katia Krafft)

that had covered the area in 1845 killing about 1,000 people, and that renewed earthquakes and small eruptions from Ruiz posed the same threat again. No coordinated plans or actions were taken, and 22,000 people were killed by the November 13 mudflow (Fig. 11.7).

The opposite situation occurred on Guadeloupe Island in the Caribbean in 1975, when La Soufrière Volcano awakened with earthquakes and small explosive eruptions. Remembering the 1902 Mont Pelée disaster on the nearby island of Martinique that killed 29,000 people, authorities evacuated 72,000 inhabitants from the city of Basse-Terre and the flanks of La Soufrière. The evacuation lasted for three months while the volcanic activity waned without a major eruption. The cost of the evacuation and the disagreements among scientists involved in monitoring La Soufrière were bitterly debated for many months afterward.

A more successful forecast was made at Colo Volcano, on a small island in Indonesia, which was shaken by an earthquake swarm and by small explosive eruptions in July 1983. On the basis of the past behavior of Colo and similar volcanoes, geologists from the Volcanological Survey

Figure 11.9. Closing the approach roads to Mount St. Helens in Washington State before the May 18, 1980, eruption stirred considerable controversy. Without these restrictions the toll of 57 deaths would have been much higher. (Photograph by U.S. Geological Survey)

of Indonesia recommended to the local government officials that the island be evacuated. The officials concurred, and all 7,000 inhabitants were evacuated by boat. A climax eruption occurred on July 23, sweeping the island with hot pyroclastic flows. The coconut plantations were destroyed (Fig. 11.8). It will take years to rehabilitate the island, but all the people survived.

The checkered history of past success and failure with eruption forecasts, and the responses to these forecasts, point out some considerations of risk and social disruption that may occur in future crises. Volcanic disasters will occur in the future, and they may be worse than the ones in the past. Since the beginning of civilization some 5,000 years ago, the world has not experienced one of the really large magnitude volcanic eruptions that we know occurred in prehistoric time. The most recent eruption that is known to have expelled more than 1,000 cubic kilometers of magma occurred at the Toba Caldera in Indonesia 75,000 years ago; in the future an eruption of this magnitude will probably occur somewhere in the world again.

The technology exists to provide warnings, although imperfect, of future volcanic eruptions over both the long-term (decades to centuries) and short-term (minutes to months) periods. These warnings will be of little value or even be counterproductive unless their implications and limitations are clearly understood by governing officials and the news media. It is essential that much of that understanding exist before the next crisis occurs. Volcanic risk can be reduced. The main problems at this time are in establishing cooperation among scientists, governing officials, and the news media who are likely to become involved in eruption crises. In addition, the public at risk must be educated about the various dangers from eruptions and how those dangers can be lessened. No one living in the shadow of a magnificent volcano wants to dwell constantly on its dangers, but it is prudent to plan for the worst and hope for the best (Fig. 11.9).

12

Volcano weather

El Chichón, Mexico: March 28, 1982

Shortly before midnight in a remote area of southern Mexico, a volcano called El Chichón suddenly blasted a huge column of gases and ash 20 kilometers into the sky. This dramatic Plinian eruption lasted for almost six hours before it waned. As it turned out, this was only the beginning of its activity. For the next few days the volcano steamed and rumbled while hundreds of earthquakes shook the region; then on April 3 a second powerful eruption sent another cloud almost as high as the first one. Some ten hours later the third and largest eruption occurred, pumping an immense cloud of volcanic gases and ash more than 25 kilometers into the stratosphere (Fig. 12.1).

The stratospheric cloud moved westward, tracked by both satellites and ground-based instruments. Within six days it was over Hawaii, where the Mauna Loa Observatory measured it to be 140 times denser than the cloud generated by the Mount St. Helens eruption two years earlier. The cloud had reached Japan by April 15, and by April 26 had circled the Earth (Fig. 12.2).

Although the Mount St. Helens eruption devastated a larger area, El Chichón's atmospheric effects were far greater. The principal difference is that Mount St. Helens erupted in a lateral blast, spending much of its energy at a low angle to the Earth's surface, whereas the El Chichón blast was aimed straight up. The Mexican volcano's magma was unusually sulfur-rich and thus produced tiny droplets, called aerosols, of sulfuric acid that augmented the cloud of fine ash in the stratosphere.

El Chichón's volcanic cloud has been the best studied in history. It was measured and sampled by ground observations, radar, high-flying aircraft, balloons and satellites; the data are still being interpreted and discussed. When all the results are in, some long-standing questions about

Figure 12.1. *The village of El Naranjo, Mexico, 9 km from El Chichón Volcano, was destroyed by pyroclastic flows. The volcano erupted violently three times between March 28 and April 4, 1982, causing some 2,000 deaths: the worst volcanic disaster in Mexico's history. (Photograph by Katia Krafft)*

the interaction of the Earth and the atmosphere may be at least partially answered.

WEATHER AND CLIMATE

The idea that volcanic dust and gases can affect weather and climate is an old one, and still the object of study by both geologists and meteorologists. Benjamin Franklin was one of the first to make the connection between the two when he suggested that the haze and cold weather in Europe during 1783–4 might have been a result of the great eruption of Laki Volcano in Iceland. Earlier in 1783, Laki had spewed forth 12 cubic kilometers of basalt, largely in effusive lava flows. A major explosive eruption of Asama Volcano in Japan took place in the same year and was probably large enough to affect worldwide climate, but almost nothing was known of that event at the time; Japan was a closed empire in the eighteenth century, with little outside communication. Not until the

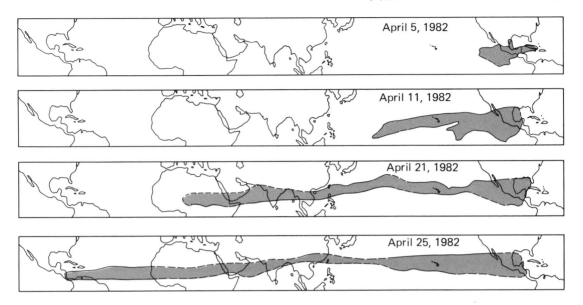

Figure 12.2. The stratospheric cloud of volcanic dust and gas from El Chichón spread westward, circling the globe in three weeks. (After Rampino and Self, Scientific American *250, no. 1, [1984]: 54)*

past century have worldwide weather records and communication been sufficient to define the overall climate of the Earth.

The words *climate* and *weather* are sometimes used synonymously, but in reality have distinctly different meanings. Weather is the momentary state of the atmosphere in terms of hot or cold, wet or dry, calm or windy; climate is the sum total of meteorological elements that characterize the atmosphere, both average and extreme, at any one place over a period of years. Volcanic activity is only one factor among many that may alter weather and climate, so it is difficult to assess its precise role. Variables such as ocean circulation, world wind patterns, carbon dioxide and other minor gases in the atmosphere, sun spots, giant meteor impacts, and systematic changes in the position of the sun and Earth also affect weather and climate. The sum total of these factors, and probably others, make the changing weather and short-term variations in climate a frequent topic of conversation.

When a volcano erupts explosively, sending a large ash cloud into the atmosphere, the immediate effect on local weather is clear and dramatic: an enveloping darkness, interrupted by lightning bolts, blocks the sun's warming rays. The rapid ascent of the cloud causes violent and erratic winds, and moisture contained in the uprushing ash, dust, gases, and surrounding air condenses into rainfall. Sometimes rain from a volcanic cloud is so laden with ash and dust particles that it falls as tiny mudballs

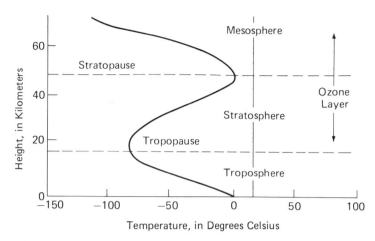

Figure 12.3. Profile of the Earth's lower atmosphere. The reversal in temperature at the tropopause acts as a lid to most of the Earth's cloud systems.

rather than raindrops. Most of these local effects occur close to the volcano, but in a very large eruption, dark ash clouds can spread downwind for hundreds of kilometers; if winds aloft are variable, "downwind" can be in more than one direction. These local to regional weather effects can be violent and disruptive, but they generally do not last long enough to have a significant effect on climate.

STRATOSPHERIC DUST AND AEROSOLS

More subtle but also more profound is the effect of volcanic dust and gases from large explosive eruptions whose clouds reach high altitudes. From the Earth's surface to about 15 kilometers, the air becomes colder with increasing altitude until it reaches about −50° C. At that level in the atmosphere, called the tropopause, the decrease in temperature reverses, and above the tropopause temperatures begin to increase slightly with altitude (Fig. 12.3). This surface, called an inversion, acts as a barrier to all but the most rapidly ascending thunderstorm or volcanic eruption clouds. It forms a lid to the Earth's "weather" in terms of cloud cover and rainfall.

Above the tropopause, the exact height of which varies a few kilometers in altitude with the seasons and latitude, is the region called the "stratosphere" – a cold, dry, eternally sunny zone of thin air stirred by high-speed winds called "jet streams." Volcanic dust and gases injected into the stratosphere by high eruption clouds – and some, like the cloud from the Krakatau eruption in 1883, can reach 50 kilometers in altitude

– are such tiny particles that they stay suspended for months to years. Because there is no rainfall to wash them back to Earth, the stratospheric winds sweep them into veils that cover a whole hemisphere or even the entire globe. Water droplets and other gases like carbon dioxide in a high eruption cloud evaporate or diffuse into the thin air. Larger dust particles settle back down below the tropopause and are washed out by rainfall. Sulfur dioxide (SO_2) gas, however, undergoes an interesting transformation. It picks up oxygen and water and forms an aerosol of sulfuric acid. These droplets and fine volcanic dust particles, most of them about one micron (0.001 millimeter) in diameter, form thin haze layers in the stratosphere that may persist for several years. Besides producing unusual visual effects such as colorful sunsets and halos around the sun and moon, these haze layers disturb the radiation heat balance between the sun, the Earth, and the cold night sky.

VISUAL EFFECTS

Well before stratospheric balloons and high-altitude jets could sample the upper atmosphere, careful observers watching the long afterglow of sunsets in the months following the 1883 Krakatau eruption were able to determine that high altitude haze layers had encircled the globe. One of the best descriptions of the spectacular sunsets seen for many months after that eruption was written by an observer in Hawaii, the Reverend Sereno Bishop:

> ... the horizon where the sun has just set is occupied by a bright silvery luster. Above this a yellowish haze fills the western sky.... This haze rapidly changes in color and extent, ranging through greenish yellow and olive to orange and deep scarlet. As the dusk advances, orange and olive tints flush out on all sides of the sky, especially in the east. The chief body of color gathers and deepens over the sunset, deep scarlet has overpowered all other hues.... There is a dark interval above the red. The stars begin to appear. While yet the color flames low, above the dark space appears a repetition of the orange and olive hues. ... Again the colors change and deepen into red, and after the stars are all out, and the earlier flame has sunk below the horizon, a vast blood-red sheet covers the west.... I have known our usual 30 minutes of twilight to be prolonged to 90, before the last glow disappeared.*

* T. Simkin and R. S. Fiske, *Krakatau* (Washington, D. C.: Smithsonian Institution Press, 1983).

High-altitude haze layers produce these remarkable optical effects by reflecting, refracting, absorbing, and scattering the rays from the sun, which, although recently set, is still illuminating these dust and aerosol veils at elevations of 20 to 30 kilometers above sea level.

The Reverend Bishop was also the first to describe the broad opalescent corona that formed around the sun after the Krakatau eruption. The phenomenon, still known as "Bishop's ring," has since been observed after large eruptions like those of Mont Pelée in 1902, Agung in 1963, and El Chichón in 1982.

COOLING EFFECTS

Absorption of the sun's shorter-length light rays by these haze layers heats the stratosphere and reduces the warming effect of solar radiation on the Earth's surface. At night, heat loss from the Earth's surface to the dark sky by longer wavelength infrared radiation is not impeded by the haze layers. To explain this wavelength dependent effect, imagine a lake covered with floating logs. Water waves whose separation is smaller than the logs will be quickly damped as their energy is absorbed by hitting the floating objects, whereas long wavelength water waves will cause the logs to rise and fall in unison and the waves will continue to move across the lake surface.

Most meteorologists studying the effects of volcanic dust and aerosols in the stratosphere agree that the normal heat-radiation balance of the Earth is disturbed; the arguments revolve around just how much. The explosive eruption of Indonesia's Tambora Volcano in April 1815 (see Chapter 10) was one of the largest in historical time (Fig. 12.4), and in 1816 the northeastern United States experienced such cold weather all year that 1816 was often called "the year without a summer." Woods Hole researchers Henry and Elizabeth Stommel have carefully studied the possible relationship of the Tambora eruption to the cold weather of 1816, and they conclude that the eruption did depress world temperatures. This phenomenon was superimposed on a generally cooler period of world temperatures that began about 1780 and lasted for several decades. They note that at many of the places where long records of temperature had been kept, the abnormal weather of 1816 was the coldest ever recorded.

As seen in Figure 12.5, the average amount of cooling following some major volcanic eruptions, including the 1815 Tambora blast, is about 0.3° C. Eruptions that appear to have the most impact on world climate are those with the following characteristics: those that produce very high

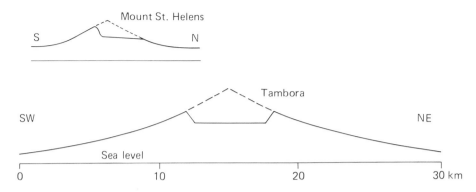

Figure 12.4. The eruption of Mount St. Helens in 1980 immediately followed a massive 3-cubic-km avalanche from the north side of the summit. The ashfall and pyroclastic flows totalled about 1 cubic km. In contrast, the 1815 eruption of Tambora Volcano in Indonesia disgorged about 100 cubic km of ash and pyroclastic flows. These true-scale profiles compare the changes in topography caused by these two explosive eruptions.

eruption clouds, those that emit large amounts of sulfur gases, and those from volcanoes located at low latitudes where their dust may circulate into both hemispheres.

Much larger eruptions than any recorded are known to have occurred in prehistoric times, and it is logical to ask if these great eruptions produced correspondingly larger climatic effects. If the stratospheric dust and aerosol veils were proportionately denser and more opaque, the answer would be yes. However, greater density of particles and droplets might have brought about their agglomeration into larger grains that would have settled out of the stratosphere more rapidly. In this case the cooling effect from a great eruption would be larger, but not as large as if the effect were directly proportional to the size of the eruption.

A long period of time with many eruptions similar to those at El Chichón, rather than isolated great eruptions like the huge prehistoric Yellowstone eruptions, may produce a more profound cooling effect. Many scholars of the ice ages think that a worldwide cooling of 6° C would bring about another major period of glaciation. Even more extreme cooling and darkness might cause major extinctions of lifeforms, like the sudden disappearance of the dinosaurs 65 million years ago.

EXIT DINOSAURS

A major scientific debate is currently in progress over the drastic change in life on Earth that occurred at the end of the Mesozoic (middle life) era. One group believes a huge meteor or comet, or a group of them, struck the Earth and produced a great dust veil that plunged the world

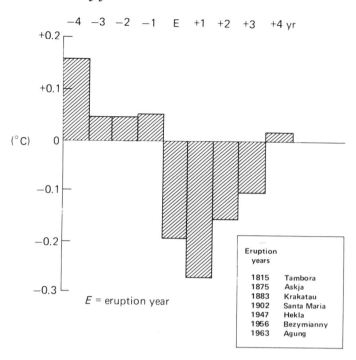

Figure 12.5. Plot of the average temperature changes in the northern hemisphere during the four years before and four years after some large explosive eruptions. The average departure from normal following the eruptions is not large, but the maximum cooling during the first year after the eruptions, and the three- to four-year period of cooling are similar to the buildup and clearing of volcanic haze caused by the El Chichón eruption in 1982 (see Figure 12.7). (After Self et al., Journal of Volcanology and Geothermal Research *11 [1981]: 41–60)*

into freezing darkness for many weeks or months. This impact hypothesis is supported by the discovery of a layer of clay highly enriched with iridium that was discovered in many sedimentary rocks formed at or near the time of extinction. Iridium is a rare element in the Earth's surface rocks but is often present in metallic meteors.

A splinter group of the impact supporters believe that sudden warming of the Earth, rather than cooling, resulted from the meteor or comet collision. In their version the impact occurred in a region with thick limestone deposits and great quantities of carbon dioxide were released into the atmosphere, resulting in a major warming effect. Carbon dioxide gas in the atmosphere transmits short-wavelength light waves incoming from the sun, but blocks the outgoing longer wavelength infrared radiation from the cooling Earth at night (Fig. 12.6). The analogy of carbon dioxide in the atmosphere to windows in a greenhouse has been labeled the "greenhouse effect."

Still another group argues that increased volcanism from sources deep

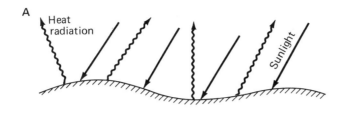

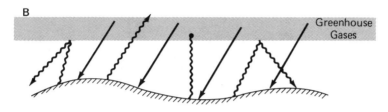

Figure 12.6. Cartoon of the greenhouse effect and how it could warm the Earth's climate. A: Incoming sunlight warming the Earth's surface is normally balanced by infrared heat loss to space. B: Greenhouse gases absorb and backscatter some of the infrared radiation, thereby raising the Earth's surface temperature.

within the Earth could also have produced iridium anomalies and the cold, dark climate for extinction. Possibly both groups are right. Perhaps it takes a one-two punch, both rampant volcanism and bombardment from space, to disturb the world enough to destroy its dominant life forms.

OZONE LAYER

Returning to more immediate problems, recent observations of the high atmosphere show alarming decreases in ozone, the gas molecule of three oxygen atoms, which protects the Earth's surface from much of the sun's ultraviolet rays. This has been largely blamed on chlorofluorocarbons (like the refrigerator gas Freon), man-made gases extensively used in air-conditioning, aerosol spray cans, and plastic foams. Chlorofluorocarbons are very stable gases, but they destroy ozone molecules. The chlorine atoms in the fluorocarbons are apparently the major trigger in this break-down reaction.

Volcanic gases also contain chlorine, and it is not yet known if this natural source of atmospheric contamination perturbs the ozone layer as well. One thing is clear: The destruction of the ozone layer is a potential problem of great magnitude to the health of most living things on Earth, so much so that the major manufacturer of freon gas has stopped making some chlorofluorocarbons, and an international treaty now calls for limits on their production. Research on all aspects of the formation, stability, and destruction of ozone in the atmosphere is urgently needed.

EL CHICHÓN AND EL NIÑO

The eruption of El Chichón was of moderate explosive magnitude, but nevertheless it lofted unusual amounts of sulfur gases and sulfuric acid aerosols into the stratosphere. Sedimentary rocks underlie the El Chichón region, and some of these strata are composed of sulfate deposits formed from the evaporation of ancient seas. Some geologists believe this was the source of the excess sulfur gases.

In the summer and fall of 1982, following the eruption of El Chichón, a major disturbance of the ocean temperature and currents in the equatorial Pacific – a phenomenon called "El Niño" – took place. Drought in Australia, heavy rains and waves on the California coast, and other weather extremes in 1982–3 were attributed by meteorologists to the major El Niño of that period.

There may be a connection between El Chichón and El Niño, and although many oceanographers and meteorologists do not agree, some find it an intriguing possibility. Changes in solar heat reaching the Earth's surface might affect the trade winds, which are less strong during El Niños, and this in turn could reduce the westward current in the equatorial Pacific.

Recent evidence of earthquake swarms and large submarine lava flows near the equatorial portion of the East Pacific Rise also suggest a linkage between volcanic eruptions and El Niños. Earthquake swarms preceded the past few El Niños, and the large and very young-appearing lava flows – up to 15 km^3 in estimated volume – are thought to have erupted on the seafloor during or just after the earthquake swarms. The enormous amount of heat in these flows could be a factor in the increasing in equatorial surface seawater temperatures associated with El Niños.

In all these connections or coincidences there are strings of "might," "possibly," and "maybe." The interactions of the solid Earth, its atmosphere, its oceans and its biosphere, and the incoming and outgoing thermal radiation are poorly understood. The exchanges among them are not only complex but are generally investigated by scientists from many differing fields, who may not understand the problems of another discipline. It is like the blind men describing the elephant, with the additional problem that each blind man speaks a different language.

The problem is a long way from being solved, but the global obser-

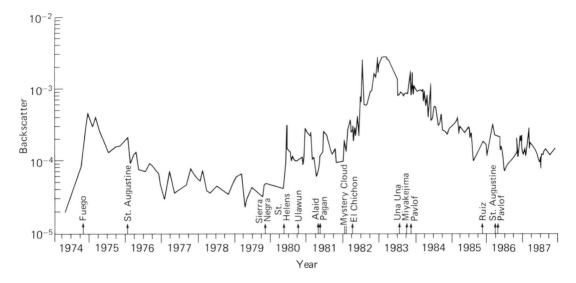

Figure 12.7. Stratospheric haze measured by LIDAR backscattering at Hampton, Virginia. Note the major buildup following the El Chichón eruption in Mexico in 1982. (From William Fuller, NASA, Smithsonian Scientific Event Alert Network Bulletin *13, no. 3 [1988]: 14)*

vations are being made and the questions are being asked. Some answers will be forthcoming in the next few decades. All the observations are contributing to a better understanding of the gases, dust, and aerosols in the stratosphere, and the chemical and physical changes taking place among them.

PROBING THE STRATOSPHERE

Among the more important new observations of the stratosphere are direct sampling by high-flying balloons and aircraft, LIDAR measurements from the ground, and observations from spacecraft orbiting high above the Earth's atmosphere. Some of the direct sampling methods collect both gases and particles at varying altitudes, while others filter the rarefied air to concentrate the dust and aerosol particles.

LIDAR uses a pulsed laser beam pointing up to the sky where some of the laser light is scattered back by haze layers in the stratosphere (Fig. 12.7). These "echoes" of light are picked up by a sensitive receiver, and the opacity and altitude of the haze layers are determined from the intensity and lag time of the backscattered laser light. LIDAR stations at scattered locations in the northern hemisphere have established that eruptions like that of El Chichón do produce a significant stratospheric haze layer over the entire hemisphere. The El Chichón dust and aerosol layer spread rapidly during a period of several weeks and diminished slowly over a period of two to three years.

Images from satellites can actually record the spread of the dense volcanic dust and aerosol layers, and special instruments can measure ozone and sulfur dioxide concentrations in the stratosphere by how much certain wavelengths of ultraviolet light from the sun are reflected back to space.

SURVIVAL

The geologic record clearly indicates that there have been dramatic changes in climate in the past, and future changes could bring about profound problems for humankind. Warming of the climate would cause a major rise in sea level, while colder climate could seriously affect food supplies. Life is tenacious; it has survived for billions of years, but individuals and species are fragile and depend on a stable and benign environment. The role of volcanism on weather and climate may be more than a scientific curiosity; it could affect the comfort and welfare of the entire Earth.

In a more immediate sense, volcanic processes have formed, and are still forming, valuable mineral deposits and geothermal power resources. These tangible assets are discussed in Chapter 13.

13

Volcanic assets

Cripple Creek, Colorado: November 25, 1914

The mine manager's voice was hushed and strained as he led the two men up the gully to the entrance of the Cresson mine. While they walked, manager Dick Roelofs explained that "something unmentionable" had happened on the twelfth level of the Cresson, an event so important, he didn't feel he could take the responsibility alone. He had asked lawyer Hildreth Frost and executive Ed De LaVergne to go down into the mine with him as his witnesses, but they were unprepared for what they saw. Carrying magnesium flares,

> ...they descended the cage to the twelfth level. Dick walked them around a half mile down there, so they wouldn't know exactly where they were. They turned off the drift into a lateral and brought up against a double steel door. Dick banged a signal against the door and it opened. Behind it were three guards armed with six revolvers. Beyond the guards at one side of the lateral was a sort of ladder-platform beneath a large hole in the wall five feet wide. Dick motioned to Ed and Hildreth to climb on the platform and stand in the hole. When the three men were lined up there Dick struck a kitchen match and lighted the magnesium flares. Ed thrust his flare through the hole into the darkness.
>
> What the three men saw stunned them as a child is stunned by his first Christmas tree. It was a cave of sparkling jewels. The brightness blinded them at first but then they made out that the jewels were millions of gold crystals – sylvanite and calaverite. Spattered everywhere among the crystals were glowing flakes of pure gold as big as thumbnails. The cave was forty feet high, twenty feet long and fifteen feet wide. Small boulders glittered on the rough floor. Piles of white quartz sand glowed like spun glass.

...The cave, [Ed] explained, was called a "vug." Technically it was a geode – a hollow, rounded nodule of rock lined with gold crystals. Ed had never seen or heard of a vug approaching the size of this one. ...During the next month Dick Roelofs' crew scraped 1400 sacks of crystals and flakes from the walls of the vug and sold them for $378,000. A thousand more sacks of lower-grade ore brought $90,637. Before Christmas, the crew stoped out the vug to a depth of several yards. This outer section realized some $700,000. Altogether, the Cresson vug produced $1,200,000 in four weeks.*

At today's gold price that would equal $25 million, making the "vug" in the Cresson mine one of the richest ore bodies ever discovered. A deposit like this one forms when gold precipitates from underground water that is circulating through the cavity. The mining district at Cripple Creek, where the Cresson mine is located, lies in an eroded and extinct volcanic caldera.

Many lodes of gold, silver, and other important though less precious metals were deposited by circulating waters heated by magma, but these are not the only benefits that can come from volcanism. Some hot groundwater systems are still active and can be tapped for geothermal power. Geysers and hot springs, like those in Yellowstone National Park, are uniquely interesting features, and most of them owe their existence to the still-hot volcanic rocks that form their roots. Soils around live volcanoes are exceptionally rich because their nutrients are replenished periodically by showers of volcanic ash; and over geologic time much of the Earth's air, water, and rocky crust are recycled and renewed by volcanic action.

An intangible asset but one of great value nonetheless is the sheer beauty of volcanic mountains (Fig. 13.1). The soaring, snow-covered majesty of Washington's Mount Rainier, California's Mount Shasta, or Japan's Mount Fuji is beyond price, and one of the world's most beautiful lakes, Crater Lake in Oregon, is cradled in the 7,000-year-old caldera of an ancient volcano.

THERMAL WATERS

Most hot springs and geysers are closely related to volcanic activity or shallow intrusions of magma beneath the Earth's surface. In these lo-

---

* Marshall Sprague, *Money Mountain* (Boston: Little, Brown, 1953).

Figure 13.1. The perfect cone of Kronotzky in Kamchatka is a prime candidate for the world's most beautiful volcano. This 3,528-meter-high stratovolcano rises 3,100 meters above the frozen countryside. It last erupted in 1922–23. (Photograph from the Institute of Volcanology, Kamchatka, USSR)

cations groundwater is heated when it comes into close proximity with magma; as it heats it becomes less dense, and circulates upward. If fractures in the rock extend to the surface, this upward leakage may form hot springs and, more rarely, geysers.

Geysers are unusual features that occur in only a few regions of the world, in such widely divergent places as Iceland, New Zealand, and Yellowstone National Park. They were first described in Iceland, where

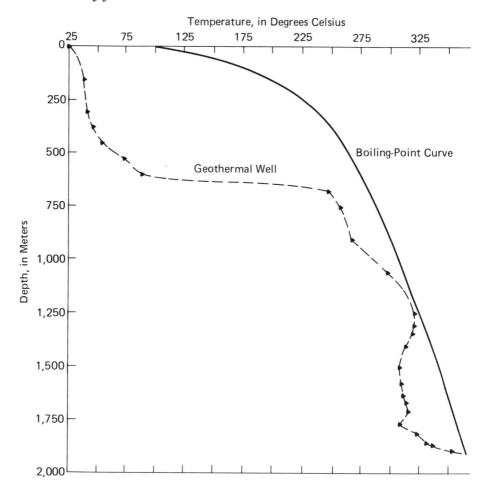

Figure 13.2. Geothermal energy is available in groundwater heated by volcanic activity. The dashed line is the temperature in a geothermal well drilled on the east rift zone of Kilauea Volcano in 1976. The solid line is the boiling temperature of water versus depth underground. Groundwater heated above 100° C flashes into steam as pressure in the borehole is reduced to near-surface conditions. (Modified from Donald Thomas, U.S. Geological Survey Professional Paper 1350, [1987]: 1512)

a spectacular geyser field is found; *geysir* is an Icelandic word meaning to gush or to rage.

A geyser is a boiling spring that intermittently spews a column of steam and hot water into the air until the conduit system that feeds it is emptied. Cooler groundwater moves in to refill the conduits and must be reheated to near-boiling temperatures before the action is repeated. Critical to the process is the fact that the boiling point of water increases with the increasing pressure deep beneath the Earth's surface. Although it is nearly 100° C at the surface, water's boiling temperature increases to 150° C at a depth of 75 meters, 260° C at 500 meters, and 300° C at 1,000 meters (Fig. 13.2).

As the surface water in a geyser begins to boil it becomes less dense, thus reducing the pressure in the conduits below. Water that has been heated to near-boiling temperatures at depth in the conduits, when subjected to lower pressure, suddenly reaches its boiling point. This deep water flashes into steam, pushing hot water and steam up the conduits and out the geyser, thus reducing pressure at depth even more. This chain reaction continues until the conduits have been discharged; then cooler groundwater seeps back in to replenish the system for the next gush of nature's percolator.

The height to which a geyser can spout ranges from less than a meter to, in rare cases, as much as 100 meters. Some geysers erupt every few minutes, while others may wait for years. For most the cycle time is irregular; even Yellowstone's Old Faithful, famous for erupting on schedule, has a repose time that varies from 40 to 80 minutes with an average of about 65 minutes.

Yellowstone National Park in Wyoming has been the site of major volcanic activity for at least the past two million years. Three great eruptions and countless smaller ones have occurred during that prehistoric period. Fortunately, Yellowstone's present activity is hydrothermal; magma is cooling underground and supplying heat for the scenic geysers and hot springs. Volcanic eruptions will undoubtedly occur again at Yellowstone, but human time is so brief compared to Earth's sandglass that there is only a small possibility of our generation seeing such a spectacle.

ORE DEPOSITS

Regions of hot, circulating groundwater are called "hydrothermal systems" by geologists. In many ways, they operate like giant stills. As molten rock underground crystallizes, volcanic gases and trace elements in the cooling magma are expelled. These compounds are dissolved in the circulating hot water of the hydrothermal system along with soluble elements leached from the porous and fractured rocks in which the system operates. Those elements in solution may be gold, silver, mercury, copper, lead, and zinc, along with many others. They are transported through pores and cracks in the rocks to sites where conditions allow them to precipitate and concentrate. Such conditions include cooling, reduction of pressure, and chemical reactions with the host rocks through which the hot solutions are moving. Here grain by grain, crystal by crystal, the

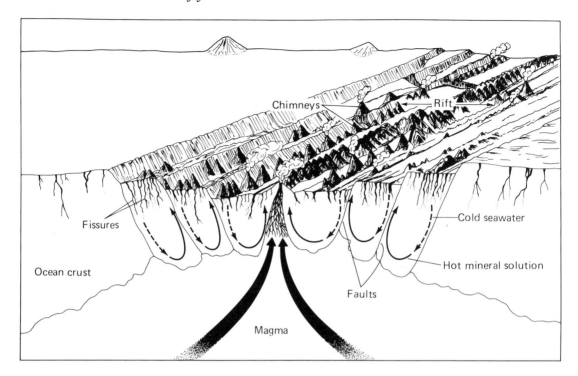

Figure 13.3. Model to explain submarine ore deposition envisions seawater circulating by convection in the fractured seafloor. Heat from magma intruding along the rift axis drives the convection and supplies some of the chemical elements. Other chemical elements are leached from the bedrock of the seafloor. Chimneys of zinc, iron, and copper sulfides are formed where the submarine hot springs are suddenly chilled by the cold ocean water. (Modified from Charles Petit, "Neptune's Forge," Science 83 [January–February 1983]: 63)

valuable minerals slowly accumulate over long periods of time. Magma supplies the heat, ground water the transportation, and specially receptive places underground the sites for deposition of valuable minerals. The art of prospecting is to find these special places.

Submarine volcanism also produces ore deposits, the potential volume of which has only recently become apparent (Fig. 13.3). In 1963 ocean-ographers discovered that some deep basins in the Red Sea contained hot brines and bottom sediments rich in zinc and copper sulfides. The Red Sea is a young volcanic rift zone, and the hot brines, rich in dissolved metals, apparently derive from ocean water as it circulates through the cooling volcanic rocks and magma that form the new ocean crust beneath the axis of the Red Sea.

In 1977 hot springs were discovered along a rifting submarine ridge in the eastern Pacific, 320 kilometers northeast of the Galapagos Islands.

Figure 13.4. Alvin, a submersible vehicle operated by the Woods Hole Oceanographic Institution, has been the discovery vessel in many of the dives to the submarine hot springs and their strange oases of seafloor life. Alvin is seen here being lowered into the ocean from the oceanographic ship Atlantis I. (Photograph by Woods Hole Oceanographic Institution)

Oceanographers in the research submarine Alvin (Fig. 13.4), diving to a depth of almost 3 kilometers, encountered undersea hot springs surrounded by colonies of strange new life forms unknown until that time. On further exploratory dives they obtained samples of the hot water, minerals, and living creatures from this and similar strange oases that are fed by thermal springs along the mid-ocean ridges. At 21° north on the East Pacific Rise, in 2,500-meter-deep water near the mouth of the Gulf of California, research submariners found chimneys of iron, zinc, and copper sulfides that were pouring out 350° C hot water laden with metal sulfide particles; they named these curious vents "black smokers. (Fig. 13.5)"

Ore deposits found at these submarine hot springs form crusts and layers near their sources on the rifting ocean ridges. As the seafloor spreads slowly away from the ridge, the lateral extent of the metallic

Figure 13.5. Chimney of metal sulfide minerals built by a submarine hot spring on the Mid-Atlantic Ridge at a depth of 3,600 meters. Vent shrimp, 2.5 to 5 cm long, swarm in the warmer water near the chimney. (Photograph by NOAA)

sulfide deposits is increased; as spreading continues the deposits are carried long distances across the ocean bottom, eventually reaching a subduction zone. There, one of two possible things occurs: Either the oceanic crust and its piggyback ore deposit are consumed in the oceanic trench, or a large slab of the crust and ore deposit may be sliced off like a wood shaving in a carpenter's plane and thrust up onto the adjacent continent.

The famous copper mines of Cyprus occur in one of these overthrust slabs. Those rich mines supplied ore for the Bronze Age of Mediterranean civilization and are still producing today. Their deposits of copper sulfide minerals are in pillow basalts and related deep-sea volcanic rocks, so they have long been a mystery to geologists, who until recently had no information on undersea volcanism. Ancient submarine hot springs and slices of old Mediterranean seafloor that were uplifted as Africa pushed into Europe now provide a reasonable explanation of these valuable mines.

Most ore deposits originating at submarine hot springs are probably consumed into subduction zones instead of being uplifted onto land. This may not be their end, however; magma rising to form volcanoes above subduction zones may be enriched by these valuable elements. This process would provide additional metals for transport and deposition into the more conventional hydrothermal ore deposits in the ruins of ancient continental volcanoes.

Thirty years ago no one would have guessed that oceanographers would one day unveil some entirely new concepts about the formation of ore deposits. It proves once again the old adage that gold is where you find it.

Diamonds also owe their origin to volcanic processes, but unlike hydrothermal mineral deposits, they form from carbon atoms that are squeezed into tight networks at great depths – perhaps 200 kilometers. It is possible that diamonds are not uncommon at those depths, but they are certainly inaccessible. Diamonds and the strange volcanic rocks in which they occur were erupted up from those great depths in rapidly formed volcanic conduits called "diamond pipes," reaching the near-surface deposits where they are mined. The reduction in pressure and temperature must have happened rapidly; if the process had happened slowly, the diamonds would have had time to transform into relatively worthless graphite, the crystalline form of carbon more common near the Earth's surface.

It is uncertain if any present-day volcanoes have the type of conduits that contain diamonds such as those that have been found in eroded volcanic pipes in Africa, Siberia, and Australia. The geologic ages of diamond-bearing pipes seem to cluster at a few selected intervals of Earth history. No one knows if the special conditions that led to their formation exist today; a complete explanation of the genesis of diamonds remains elusive and mysterious.

GEOTHERMAL POWER

The internal heat of the Earth is an enormous store of energy, and has great potential to be a significant source of power. Power is energy used in some specific unit of time, and this distinction between energy and power is critical to the development of geothermal resources. For example, the heat flow out of the Earth's surface averages only 0.06 watts per square meter. Somehow that power has to be collected from a large area to make its use economically feasible. The occurrence of underground hot water and steam that can be tapped by a drill hole is one answer to this enigma.

In most nonvolcanic areas the increase in temperature with depth is small, and underground water hot enough for efficient use is likely to occur only at depths of 4 kilometers or more. This is too deep for the cost of drilling to be recovered from use of the potential power. In volcanic areas, where slowly cooling, shallow magma bodies occur, ground water may be heated to 200° C or 300° C at depths of only 1 or 2 kilometers. These active hydrothermal systems can be tapped by boreholes and the hot water and steam used for electrical generation or heating purposes. Power from these shallower, hotter hydrothermal reservoirs can pay for the cost of drilling and of building power plants.

If the underground water is hotter than 150° C, geothermal power can be extracted using the same principle that drives a geyser – the fact that the boiling temperature of water increases with increasing depth. Geothermal wells are often drilled into these higher-temperature reservoirs with compressed air to lift out the rock chips. This open-hole drilling lowers the pressure at the bottom of the well. When encountered, the water in the hydrothermal reservoir will be well above its surface boiling temperature and will flash into steam. A mixture of steam and hot water will spurt from the drill hole, and the steam can then be separated and piped to a turbine generator plant (Fig. 13.6).

In a few geothermal fields like The Geysers near San Francisco, California, high-pressure steam rather than hot water exists in the underground reservoir. In this unusual situation the problems of separating steam from hot water do not occur, so electrical power is even easier and more economical to generate.

Lower-temperature hydrothermal systems are less economical to use for electrical generation but are still of great value for space heating homes and greenhouses, and for industrial purposes. Reykjavik, the capital of Iceland, is a city heated almost entirely with hot water piped from underground. In Iceland, a large island of volcanic origin and frequent

Figure 13.6. Steam and hot water flow from a geothermal discovery well on São Miguel Island in the Azores. The potential power output of a well is tested by measuring the flow rate of steam and water at various wellhead pressures. (Photograph by Roger Henneberger, Geotherm-Ex, Inc.)

volcanic activity, the people are reminded by their warm, comfortable homes and plentiful hot tap water that volcanoes can be beneficial as well as destructive.

There are many volcanic areas in the world where the rocks at depth are hot but do not contain much groundwater. Recent experiments at the Los Alamos National Laboratories in New Mexico indicate that if surface water is injected into deep hot rocks through a borehole and recovered in an adjacent deep well, geothermal power can be extracted from an artificially created hydrothermal system. It is not yet clear whether or not this complex technology can produce power at a cost that would be reasonable when compared to other energy sources currently available. Tapping power directly from shallow bodies of magma is also an enticing prospect, but so far the engineering problems of drilling into magma and inserting some sort of heat-extraction device have not been solved.

The potential of geothermal power is enormous. The U.S. Geological Survey has estimated that the energy stored in hydrothermal systems beneath the United States is twice the energy in the world's known oil reserves. If hot, dry rock of the type being investigated by Los Alamos National Laboratories is included, the estimate soars to 6,000 times. There is a catch, however: Much of this energy is difficult to reach and may cost more to gather than it is worth. It is analogous to the problem of a large low-grade mineral deposit. The ocean contains an enormous amount of dissolved gold, but to recover it in pure form would cost much more than its value. The key question concerning the huge store of geothermal energy is whether it can be extracted fast enough and economically enough to produce useful power. Some of it already has been put into use and more will be, but it is probably not the complete answer to the dream of a cheap, endless source of power.

ORE DEPOSITS AND GEOTHERMAL POWER

Ore deposits of hydrothermal origin are usually discovered tens of millions of years after their volcanic rocks and hot waters have cooled. In geologic time they were yesterday's geothermal fields. Many currently active geothermal fields are probably depositing valuable minerals, but they either haven't been circulating long enough to form large deposits, or they are too hot and deep to prospect for ore bodies. They in turn may be tomorrow's ore bodies.

There is at least one major exception to this situation. A large gold deposit has recently been discovered in outcrops and drill holes on Lihir Island in Papua New Guinea, but this rich deposit lies within the caldera of a live volcano and the hydrothermal system that deposited the gold is still hot and active. Both the gold and the potential geothermal field appear to be valuable resources. It is ironic that the still-active hydrothermal system will probably interfere with mining the gold.

RICH SOILS

Volcanoes are responsible, at least in part, for some of the world's most fertile soil, and no place is this more evident than in the tropics. For example, both the islands of Java and Kalimantan (Borneo) lie in tropical latitudes in the Indonesian archipelago. On Java, small farms produce two to three crops of rice per year and have for centuries. By contrast, on Kalimantan many of the farms are temporary. The jungle is slashed and burned, and crops are planted for a few years; soon the soil is exhausted and new fields must be hacked from the forest.

One of the main reasons for this difference is the presence of live volcanoes on Java; there are none on Kalimantan. In humid tropical climates, soil is rapidly leached by excessive rainfall of most soluble minerals including potassium and phosphorus, two elements essential to plant growth. The eventual result is a red soil that is composed largely of iron and aluminum oxides and is poor in plant nutrients. Active volcanoes, especially the explosive ones, renew the soil with volcanic ash. Potassium and phosphorus are present in most of the common volcanic rock types, and weathering slowly releases these elements into soluble compounds that are needed by growing plants.

Very little research has been done on the specific chemical and physical reactions that make volcanic soil so fertile. It is an area of investigation that falls between volcanology and soil science; more cooperation between these specialists could be of both scientific and practical importance. Nevertheless, there is widespread empirical evidence that new volcanic ash invigorates plant growth. Record crops of apples and wheat in Washington State were harvested for several years after the cloud from the 1980 eruption of Mount St. Helens dusted the fields with 1 to 2 centimeters of ash. Evidence of the 1912 Katmai eruption can easily be seen in the tree ring pattern of alpine firs buried by a few centimeters of ash. The 1912 and 1913 rings are small – probably from the direct trauma of the ashfall – but for the next decade the rings show that the growth was much more vigorous than usual (Fig. 13.7).

Of course, it is possible to have too much of a good thing. An ashfall thicker than about 20 centimeters will kill most vegetation, and this new ash either must be plowed into the underlying soil or must weather over a period of years before fields can recover. The fertile soils near live volcanoes benefit hundreds of millions of people. At the same time, however, volcanoes lure larger and larger populations to live within their potentially deadly reach. In Indonesian legends good and evil battle in many forms and the outcome is generally a long-term standoff; nature's lessons are both kind and cruel.

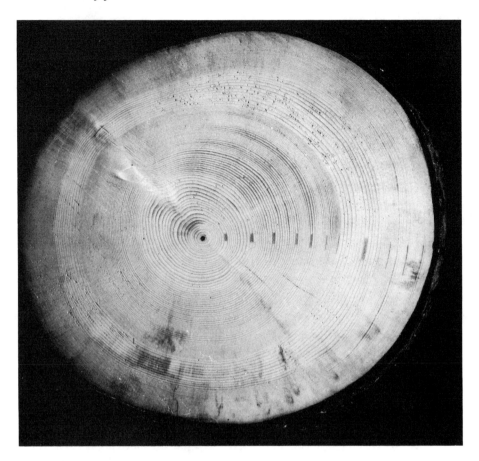

Figure 13.7. A 120-year-old white spruce tree cut in 1962 for a roadway 48 km northwest of Mount Katmai, Alaska. Each marked ring is 10 years old. Three very thin rings after 1912 are followed by 12 years of thicker than normal growth rings. Thickness of the 1912 ashfall in this area is 15 cm.

LAND, SEA, AND AIR

Thirty years ago many geologists believed that much of the Earth's air and water were exhaled from volcanoes over geologic time. Most of the gases released from volcanoes occur in just about the right proportions to account for the composition of the oceans, atmosphere, and sedimentary rocks like limestone.

Because the concept of plate tectonics has revolutionized how geologists think about Earth history, ideas about the origin of water, air, and even the continents themselves have drastically changed. The plate tectonic theory indicates that oceanic crust and much of the sediments draped on it are recycled back into the Earth's mantle. At the present rate of plate motion it takes up to 200 million years from the time new oceanic crust is formed at a mid-ocean rift until it is consumed in a subduction zone. During its long trip from ridge to trench, layers of sediment accumulate on the slowly moving seafloor. Assuming that the average rate of plate motion has been roughly the same during the 4-plus billion years of Earth history, this recycling process must have been repeated about twenty times.

In this model chemicals are added to the ocean by rivers eroding continents, and by hydrothermal systems along the mid-ocean rifts leaching the young volcanic rocks. Chemicals are subtracted by reaction of seawater with rocks in the hydrothermal systems, and by the subduction of oceanic crust and its covering sediments. For example, magnesium salts formed by the weathering of continental rocks are slowly added to the sea, carried there by rivers. If this process had continued over geologic time, the amount of magnesium salts in the ocean would greatly exceed the amounts actually present. The missing magnesium ions can be accounted for by their reaction with the hot basaltic lavas that heat and circulate submarine hot springs; the solid magnesium silicate formed by seawater circulating down into these hot lavas stays in oceanic crust and is eventually consumed in a subduction zone. The problem with the long-term, nonrecycling concept is that ocean chemistry would have slowly become more concentrated throughout geologic time. In the recycling concept, a long-term balance is maintained by subtraction of soluble ocean chemicals to offset their continual addition.

According to this latter concept, the composition of the oceans and oceanic crust are not the result of a long-term process that spans the entire history of the Earth, but instead are largely the result of the last 200 million years – the cycle time for seafloor spreading. Volcanoes fit into this grand view by recycling much of the Earth's water, carbon dioxide, and crust rather than having produced them continuously from within the Earth since its beginning.

By either concept, without the continuous formation of air, water, and land by volcanoes bringing these materials from inside the Earth to its surface, or the recycling of some of these products through seafloor spreading and subduction, the Earth and life on it as we know it would not exist. Volcanism is not the only process in planetary evolution, but it is an important one. Just how important is considered in Chapter 14, which discusses the possible role of volcanic activity on the origin and evolution of life on Earth.

14

Volcanoes and life

Eastern Pacific Ocean: February 19, 1977

As the research submarine Alvin settled slowly toward the ocean floor, 2,500 meters deep at the Galapagos Ridge axis, there was little indication that this was going to be a day of amazing discoveries. Scientists John Edmond of the Massachusetts Institute of Technology and John Corliss of Oregon State University watched through Plexiglas portholes as the submarine cruised over the gently sloping sea floor; then the pilot stopped to collect some volcanic rocks with Alvin's mechanical arm. As John Edmond* describes it:

> ... a couple of large purple sea anemones engaged our attention. Only when our gaze finally wandered away from them did we realize that the water within the range of our lights was shimmering, like the air above a hot pavement. The hastily measured temperature of the water was five degrees above the ambient water temperature (2.05° C). With all thoughts of rocks forgotten, we captured a sample of the water and then continued on our course upslope. Here we came on a fabulous scene.
>
> The typical basaltic terrain at the ridge axis is bleak indeed. Monotonous fields of brown pillows are cut by faults and fissures. One must examine several square meters to find a single organism. Yet here was an oasis. Reefs of mussels and fields of giant clams were bathed in the shimmering water, along with crabs, anemones and large pink fish. The remaining five hours of "bottom time" passed in something close to frenzy. We recorded the temperature, conductivity, pH and oxygen content of the water; we made photographs; we sampled the

---

* John M. Edmond and Karen Von Damm, "Hot Springs on the Ocean Floor," *Scientific American* 248, no. 4 (April 1983): 86. With permission.

water; we made sure that a representative selection of organisms was collected, all under the growing pressure of steadily decreasing voltages for our equipment.

Fortunately the equipment worked flawlessly. It soon became apparent that we had come on a hot-spring field. Warm water streamed from every orifice and crack in the sea floor over a circular area about 100 meters in diameter. The temperature of the water was highly variable, but the maximum was about 17° C. The organisms were quite selective. They choked the warmest vents. In some cases mussel reefs actually channeled the flow of water, forming conduits themselves. We worked until the "scientific power" ran out.

Fifteen dives were made in that Galapagos series, and a wealth of samples and data were collected. Although it had long been suspected that there might be undersea hot springs at mid-ocean ridges, this was the first confirmation of their existence and the first view of the exotic creatures nourished by them.

On dives to several other mid-ocean ridges in the past decade oceanographers have encountered similar strange creatures at submarine hot springs, suggesting that many colonies of unusual life forms dot the rifts between separating crustal plates. It is clear that these newly discovered assemblages of living creatures are nourished by the heat and sulfur from active submarine volcanoes and shallow intrusions of magma beneath the ocean floor.

Seawater circulates down into the cooling magma beneath mid-ocean ridges, is heated, and rises back to the seafloor in the form of hot springs, carrying hydrogen sulfide (H_2S), carbon dioxide (CO_2), and other volcanic gases. In addition, sulfate in the circulating seawater is reduced to sulfide in the oxygen-poor region near the magma. Bacteria living near the submarine hot springs get their energy from combining the hydrogen sulfide gas in the hot springs with oxygen from the surrounding seawater, and with this energy combine water and carbon dioxide to form organic hydrocarbons. In the same way that plants and sunlight form the basis of the food chain on the Earth's surface, these strange bacteria and volcanic heat feed the tube worms, clams, and, in turn, the crabs in these warm dark oases deep beneath the sea (Fig. 14.1).

The strange creatures that live at deep submarine hot springs are a remarkable discovery. They clearly demonstrate that some forms of life can live on volcanic power. Sustaining life is an essential function, but there are even more fundamental ways in which volcanic activity may have played important, perhaps crucial, roles in the origin and evolution of life on Earth.

Figure 14.1. Tube worms are among the strange living creatures found in submarine hot springs oases. Bacteria within the 1- to 2-meter-long worms live on heat and hydrogen sulfide from the hot water and oxygen from seawater, and provide the basis of the unusual food chain. (Photograph by Woods Hole Oceanographic Institution)

Volcanic activity is a credible mechanism for providing the range of physical and chemical environments thought favorable for the origin of life. Many hypotheses about the beginning of life have been proposed, but until biochemists can create life in a laboratory experiment the pathway between complex organic molecules and primitive living organisms will remain obscure. What is known is that life, once begun, sustains and duplicates itself in habitats where liquid water, energy, and chemical nutrients are available. It seems reasonable that these same conditions were important in the origin of life.

Volcanic activity produces a wide variety of environments, with temperatures ranging from barely warm up to 1,200° C, hot springs, and a range of concentrations of molecules containing hydrogen, oxygen, carbon, sulfur, and nitrogen. All the other chemical elements are present in volcanic rocks, and the concentrations of these elements vary widely during the formation, eruption, cooling, and weathering of volcanic products. In explosive eruption clouds static electricity on ash particles generates large lightning storms, and classic laboratory experiments have shown that electrical sparks can produce complex organic molecules when discharged through a vapor of methane, ammonia and water.

ORIGIN OF THE EARTH

The present consensus of astronomers and cosmologists is that Earth formed 4.6 to 4.5 billion years ago while our solar nebula was condensing from part of a larger interstellar cloud. The heavier chemical elements in that cloud had been forged in one or more stars that had formed and after eons had exploded, during the earlier history of our 15- to 20-billion-year-old universe. The primeval Earth was a sphere of swept-together lumps of condensed stardust.

Melting within the Earth soon began, and presumably formed magma that fed volcanoes. Heat was available from the energy of the colliding lumps and from short-lived radioactive elements. Initial melting allowed heavy elements like iron to settle toward the Earth's center while the lighter rocky materials rose toward the surface. This gravitational settling released enormous amounts of energy – enough to melt most of the rocks and metals not previously melted by collision or radioactivity.

During this period, estimated to be the first 700 million years after the Earth's accretion, volcanoes are thought to have been extremely active and widespread, producing great volumes of lava. Many geologists think that while the Earth's core and mantle were forming, an ocean of molten rock may have covered much of the Earth's surface.

ORIGIN OF LIFE

Life as we know it certainly did not originate in an ocean of molten lava, but began after this sterile Earth slowly cooled. Life has many definitions, but its most basic component is the genetic material called DNA (deoxyribonucleic acid) that allows life forms to duplicate, or nearly duplicate, themselves. Genetic engineers have been able to modify existing strands of DNA, but they have never created it. Even the DNA in the most simple bacteria is an incredibly complex ladder of amino acid molecules.

Volcanic activity probably played a major supporting role in the dawning of life on Earth by helping to form a favorable environment for life based on DNA. The Earth's early atmosphere and oceans are thought to be derived from two sources: from volcanoes exhaling steam, carbon dioxide, and other gases; and from bombardment by comets composed largely of ice. The early atmosphere was probably a mixture composed mainly of nitrogen and carbon dioxide.

By 3.8 billion years ago, Earth's pressure, temperature, and composition were apparently favorable for the accumulation and preservation of amino acid molecules, most likely in shallow ocean waters. But DNA is like a book and amino acids are only the letters that form the words. The problem is forming that first DNA; after that it can replicate and evolve on its own. Some scientists think that life originated elsewhere in the universe and was carried here, perhaps as some primitive living cells incorporated in icy comets that bombarded the infant Earth. Others believe the origin of life is a metaphysical problem and not amenable to scientific understanding. Still others believe that the first DNA formed by chance; that possibility is not zero, but it is vanishingly small.

An interesting hypothesis by a few investigators, led by A. G. Cairns-Smith of the University of Glasgow in Scotland, is that clay minerals may have played an important role in the origin of life. Minerals have ordered internal arrangements of molecules and can grow by adding new layers of molecules in these same patterns. Clay minerals have highly complex structures of layered silicate molecules, like multitiered sandwiches; they can incorporate water molecules and even organic molecules between their patterned layers.

Cairns-Smith and others believe that such complex clay-organic molecules, once formed, could replicate themselves in the same manner as more simple clay minerals – that is, by nonbiological crystal growth. The conditions needed for such crystal growth are favorable temperature, pressure, and supply of the constituent elements; many places on Earth provide these favorable conditions. For example, places where hot springs are forming bubbling mud pots and where rocks are being slowly weathered to soil.

In this clay-mineral hypothesis, the complex silicate-organic molecules were not alive but formed templates on which some primitive DNA may have assembled itself. In a way the clay-mineral hypothesis is still a life-by-chance idea, but instead of overwhelming odds against it, the clay-organic molecules help to stack the deck in favor of DNA, and life, to form.

The origin of life, whether common in the universe or unique to Earth, remains a mystery. Once it began on Earth, however, its evolution was influenced by many factors. Food and energy supplies governed the population of various organisms. Sunlight, carbon dioxide, and nutrients such as phosphorus, potassium, and nitrogen are required by photosynthetic life forms. Habitat was a major controlling factor; warm ocean water was more hospitable than the dry land that was baked by day and frozen by night. Geologic changes strongly contributed to evolution; evaporation of a lagoon could destroy the life in it, while rise in sea level might greatly expand a favorable habitat. Active volcanoes played an important role in these changing settings.

VOLCANOES AND EVOLUTION

The presence of fossils in ancient sedimentary rocks shows that evolution from simple to complex living organisms has been going on for 3.5 billion years. For most of that long history, life consisted of simple bacteria living in mats and colonies on or near the surface of shallow ocean waters. Many of these simple cells were plantlike; they lived on sunlight, water, and carbon dioxide, and produced oxygen. By 1.5 billion years ago enough oxygen had accumulated in the atmosphere to allow more complex cells to evolve. The new cells consumed oxygen and hydrocarbons, and released water and carbon dioxide. They did not depend directly on sunlight for their energy but on oxygen and hydrocarbons formed by more primitive bacteria. This second step was a major event; it led to the evolution of multicelled organisms, which spread to particular habitats throughout the seas. By 400 million years ago plants and amphibious fishes had begun to inhabit the barren land of the continents. In terms of geologic time it was not long until forests, reptiles, and mammals followed, and evolution became even more complex.

The father of the theory of evolution, Charles Darwin, formulated some of his early ideas on the subject during his visit to the volcanic Galapagos Islands in the early nineteenth century. There he noticed that different species of land tortoises inhabited individual islands. Because almost no interbreeding of land tortoises among the separated islands took place, each colony apparently evolved its own distinct characteristics.

Darwin also noted several unique species of finches in the Galapagos; some with strong beaks for cracking tough seeds, others with long, thin beaks for plucking insects from holes in tree bark. He reasoned that the early finches that arrived on these volcanic islands, perhaps blown there in a storm, found little or no competition from other birds, and that natural variations in beak size and shape slowly led to the evolution of several species of finches – each adapted to a particular food supply and habitat. Darwin did not know about DNA, but he recognized that some genetic mechanism allowed variations among individuals to occur.

In modern terms, evolution results from two major processes: the occasional wear and repair of DNA causes variations in form and function of individual living organisms, and chance and fitness decide which species will survive. Darwin viewed this natural selection as a slow, steady, gradual process. A more recent amendment of his theory holds that widespread extinctions of life caused by major geologic or extraterrestrial catastrophes provided opportunities for periods of more rapid evolution. This concept, called "punctuated evolution," suggests that during periods of stability many variations in life forms – even beneficial ones – lose

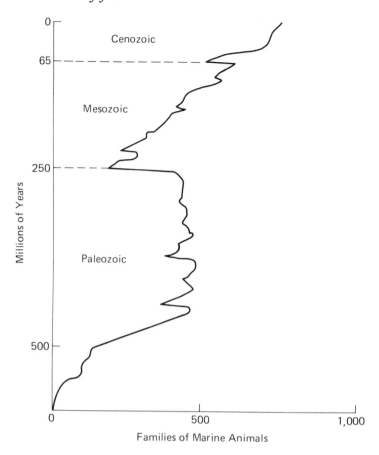

Figure 14.2. *Growth and extinctions of marine animal families during the last 600 years. The major extinctions and subsequent changes that occurred 250 million and 65 million years ago have long been recognized, and they provided the basis for dividing the fossil record into its three major eras. Recent debates about the cause of the extinctions focus on giant impacts by meteors or comets, or on extraordinary periods of volcanic activity. (Data from J. J. Sepkoski, in D. M. Raup and D. Jablonski, eds.,* Patterns and Processes in the History of Life *[Berlin: Springer-Verlag, 1986] pp. 277–96)*

out to the ecological status quo and thus slow down the evolutionary process. Climate changes or huge volcanic eruptions, however, may cause global or regional extinctions of many species (Fig. 14.2); this in turn opens up opportunities for small populations of new species to compete successfully and possibly move into unoccupied territories.

This new view, with long periods of slow evolution punctuated by rapid bursts of evolutionary change, in no way conflicts with Darwin's basic concept that mutation and natural selection are the basic driving forces of evolution. It simply adds another variable: the chance of sudden change in environment.

Darwin clearly understood that a new volcanic island provided an

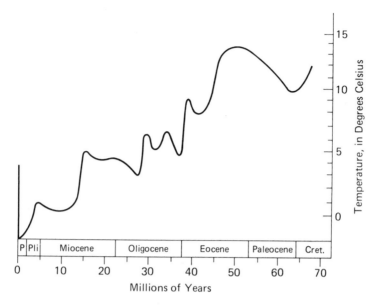

Figure 14.3. Estimated temperature at the deep sea bottom during the past 70 million years. The shorter-term dips in the long-term cooling record correspond to periods of increased explosive volcanism. (Adapted from Daniel I. Axelrod via Savin, Geological Society of America, Special paper 185, [1981]: 38)

empty territory on which immigrating plants and animals could evolve rapidly in isolation and form new species. His world view, however, was one of gradual change rather than of rapid change by catastrophes of meteoric or volcanic origin that would cause widespread extinction of many species.

Evolution might progress very slowly on a world with primitive life where the environment was extremely stable, and life might end if the environment changed too drastically. Earth with its ongoing plate tectonics, volcanism, and occasional collision with an asteroid may provide just the right challenge for rapid evolution.

The past 2 million years have seen many cycles of climate change, alternating between cold and warm, with the last glacial phase ending about 12,000 years ago. These climate changes put severe stress on plants and animals; the waxing and waning ice cover swept large continental areas clear of nearly all life, and then reopened large territories for new colonization. Coastal lowlands were drowned and then drained as ice caps melted and re-formed, with sea levels changing by more than 100 meters. Just to survive was a challenge. Humans did, and evolved rapidly in one of the great transitions in evolution. Humans have brought more change to the Earth than have most natural processes, especially in the

Figure 14.4. Footprints in the ash from the 1790 explosive eruption of Kilauea Volcano in Hawaii. Where may the longer path of time lead? (Photograph by Katia Krafft)

past 200 years, and the outcome of this latest evolutionary step is not yet clear.

The ice ages have been caused by several things. Changes in the ocean currents, brought about by the slow rearrangement of the continents, have apparently been an important factor in the cooling of the Earth during the past 50 million years. Reduced ocean circulation leads to more extreme and generally colder climates as the mixing of warm tropical water and cold arctic water is impeded. Periods of intense volcanism have accelerated that cooling trend, as explained in Chapter 12, and caused rapid drops in the worldwide temperature lasting a few million years. Volcanic eruptions do not seem to cause the off-again, on-again ice ages; that appears to be more an effect of the Earth's orbit and inclination to the sun. Volcanoes, however, do appear to have played a major role in the 10° C average cooling of the Earth's climate to the point where the glacial cycles began (Fig. 14.3).

The future of planet Earth will be influenced by many factors. Human activity will continue to modify the landscape drastically and to change the chemistry of the oceans, atmosphere, and groundwater. The risks from great volcanic eruptions will increase as human populations continue to grow (Fig. 14.4), but will be relatively small compared to the dangers of overpopulation and global war.

The Earth has survived 4.5 billion years of change, and will survive those brought about by humankind. Volcanoes will do their part in renewing the land, the oceans, and the air. In violence and in calm, the Earth abides.

Glossary

aa   A type of lava flow having a rough, jagged surface.

active volcano   A volcano that is currently erupting, or has erupted in recorded history.

aerosol   A suspension of fine liquid or solid particles in air.

airfall deposit   Volcanic debris that has fallen from an eruption cloud.

andesite   A gray volcanic rock common to stratovolcanoes, with a silica content between basalt and dacite.

ash, volcanic   Fine fragments of lava or rock, down to dust size, formed by volcanic explosions.

ash cloud   A cloud of ash formed by a volcanic eruption or a pyroclastic flow.

ashfall   Volcanic ash falling from an eruption column or ash cloud.

avalanche   A large mass of earth, rock, volcanic debris, etc., descending swiftly down a mountain.

basalt   Dark-colored lava rich in iron and magnesium, containing about 50% silica.

block, volcanic   A solid fragment of lava or rock thrown out in an explosive eruption; larger than 64 mm in size.

bomb, volcanic   A lump of lava thrown out of a volcano while still molten; takes on a rounded shape.

caldera   A large basin-shaped depression at a volcano's summit, usually formed by collapse.

cinder, volcanic   A lava fragment of about 1 cm in diameter.

cinder cone   A steep hill formed by the accumulation around a vent of cinders and other fragments expelled in an eruption.

composite cone   *See* stratovolcano.

compressional margin   The converging edges of two tectonic plates.

conduit   The crack or tube through which magma moves.

continental crust   The solid outer layers of the Earth beneath continents; less dense and thicker than oceanic crust. Normally about 25 km in thickness.

continental drift   The theory that slow relative movements of continents is caused by horizontal movements of the Earth's surface.

crater   A bowl-shaped depression around the mouth of a volcano.

crystalline rock   A rock composed of interlocking crystals.

curtain of fire   A line of lava fountains erupting along a fissure.

dacite   A light-colored volcanic rock, intermediate in silica composition between rhyolite and andesite.

dike   A blade-shaped body of intrusive igneous rock that cuts across the layering of the country rock.

directed blast   A hot mixture of rock debris, ash, and gases, generated by a volcanic explosion, that is propelled horizontally away from the vent at a high speed.

dormant volcano   A volcano that is not currently erupting but is considered likely to do so in the future.

drainback   Lava that returns underground by flowing into a fissure or back into the vent from which it erupted.

dust, volcanic   Fine particles of volcanic ash.

earthquake swarm   A closely spaced sequence of earthquakes of approximately the same magnitude; as opposed to a sequence of a strong earthquake with diminishing aftershocks.

earthquake wave   A vibrational wave produced by an earthquake.

echelon   Geologic features like faults or fissures in a sidestepping or staggered arrangement.

effusive eruption   An eruption consisting mostly of lava flows (as opposed to explosive eruption).

eruption cloud   A cloud of gas, ash, and other fragments generated by a volcanic eruption.

explosive eruption   A sudden expansion of gases laden with volcanic fragments; caused by explosive boiling.

extensional margin   The edges of tectonic plates that are moving apart.

extinct volcano   A volcano that is not expected to erupt again; a dead volcano.

fault   A fracture in the Earth's crust along which there has been movement.

fault scarp   A steep slope or cliff formed by movement along a fault.

feldspar   A light-colored mineral composed largely of oxygen, silicon, and aluminum.

fissure   A large, blade-shaped crack in the Earth.

flank eruption   An eruption that issues from the side of a volcano instead of from the summit.

flow front   The leading edge of a moving lava flow.

fumarole   An opening in the ground from which volcanic gases and steam are emitted.

fume cloud   A gaseous cloud without volcanic ash.

geophysics   The physical and mechanical aspects of geology.

geothermal energy   Energy derived from the Earth's internal heat.

geothermal gradient   The rate of temperature change with depth in the Earth.

geothermal power   Power generated by the Earth's heat energy.

granite   A coarse-grained igneous rock composed mostly of quartz and feldspar.

greenhouse effect   The trapping of solar energy by gases, such as carbon dioxide, in the atmosphere.

heat transfer   Movement of heat from one place to another.

Holocene   The period of geologic time since the last major glaciation, about 10,000 years ago to the present.

hot-spot volcanoes   Volcanoes related to a persistent heat source in the mantle.

hydrothermal reservoir   An underground zone of porous rock that contains hot water.

igneous rock   Magma or lava that has cooled and solidified, below or above ground.

ignimbrite   The widespread deposit left by a large pyroclastic flow.

intrusion   A rock body formed by magma forcing its way into surrounding host rock and then cooling; also the process of forming such a rock body.

island arc   A curving chain of volcanic islands, formed at compressional plate margins.

kipuka   An area of vegetation surrounded by a lava flow.

lava   Magma that has reached the Earth's surface; also the resulting rock when cooled.

lava channel   The faster-moving, more incandescent portion of an active lava flow, or its solidified remains.

lava dome   A steep-sided mass of viscous lava, usually with a rounded top, extruded from and covering a volcanic vent.

lava flow   A stream of molten rock, usually erupted nonexplosively, that moves downslope from the vent.

lava fountain   A jet of incandescent lava sprayed from a vent by the rapid expansion of volcanic gases.

lava lake   A lake of molten lava in a volcanic crater or depression; also solidified or partly solidified stages of the lava lake.

lava tube   A tunnel beneath the surface of a solidified lava flow, formed when the surface cools and subsurface molten rock is still flowing. Also the cave formed by emptying of the tunnel as the eruption stops or shifts.

linear vent   A long vent that forms along a fissure (as opposed to a single crater).

low-velocity layer   The zone in the upper mantle where seismic velocities and strength are lower than in the layers above; about 60 to 250 km below the surface.

magma   Molten rock within the Earth; magma that reaches the surface is called "lava."

magma chamber   An underground reservoir in which magma is stored.

magmatic gases   Gases such as water, carbon dioxide, and hydrogen sulfide, dissolved in magma.

magnetic field   A spacial pattern of magnetic forces generated by a magnetic body.

mantle   The zone of the Earth below the crust to a depth of 3,480 km; above the core.

microearthquake   An earthquake too small to be felt but detectable with a seismograph.

mudflow   A water-saturated mixture of mud and debris that flows downslope under the force of gravity.

neck   A pipelike vertical intrusion, usually seen as an erosional remnant, that represents a former volcanic conduit.

normal fault   An inclined fault in which the upper block moves relatively downward.

nuée ardente   A dense "glowing cloud" of hot volcanic ash and gas erupted from a volcano; moves rapidly downslope; a pyroclastic flow.

obsidian   A black or dark-colored volcanic glass generally rhyolitic in composition.

oceanic crust   The Earth's crust where it underlies the oceans, without the granitic layer that forms continents. Normally about 5 km in thickness.

olivine   An olive-green mineral composed of iron, magnesium, silicon, and oxygen.

ore   Any rock material containing constituents from which minerals of commercial value can be extracted.

ozone layer The layer in the high atmosphere containing ozone, the gas molecule of 3 oxygen atoms, that protects the Earth's surface from much of the sun's ultraviolet rays.

pahoehoe A basaltic lava flow with a smooth, billowy, or ropy surface.

partial crystallization The stage of cooling magma when it is partly solid crystals and partly liquid rock.

partial melt The stage of melting rock when it is partly liquid rock and partly solid crystals.

Pele's hair Strands of natural spun glass generally formed in lava fountains.

pillow lava Rounded, sacklike bodies of lava that form underwater.

pit crater A crater formed where material has been withdrawn from below and the surface has sunk.

plate tectonics The theory that the Earth's crust is broken into about a dozen plates that move slowly in relation to one another.

plume A column of hot, plastic rock rising from deep within the mantle to form hot-spot volcanoes.

pluton A large igneous intrusion that cools and solidifies beneath the Earth's surface.

precipitate A solid forming from a solution.

pressure ridge An uplift of an area of hardening crust on a lava flow, probably caused by pressure of lava still flowing beneath it.

pumice A form of volcanic glass so filled with gas bubbles and holes that it resembles a sponge and is very light.

pyroclastic deposit The deposit of volcanic fragments from a pyroclastic flow.

pyroclastic flow *See nuée ardente.*

quartz A rock-forming mineral composed of silicon and oxygen.

radioactivity The emission of radiation and/or atomic particles from atoms that are changing their composition.

repose time The interval between eruptions of an active volcano.

ridge, oceanic A major submarine mountain range.

rift system The oceanic ridges, more than 60,000 km long, where plates are separating and new crust is being created; also their on-land counterparts like the East African Rift.

rift volcano A volcano located along the rift system.

rift zone A region of cracking and pulling apart.

Ring of Fire The region of converging plate margins, with the resulting volcanoes and earthquakes, that surrounds the Pacific Ocean.

rhyolite   A fine-grained volcanic rock with a high silica composition; similar to granite.

seafloor spreading   The aspect of plate tectonics that concerns the creation of new seafloor at the oceanic ridges as the plates separate.

seamount   An isolated submarine mountain, usually volcanic.

seismic wave   *See* earthquake wave.

seismograph   An instrument that records seismic waves in the Earth's crust.

seismology   The study of seismic waves, earthquakes, and the Earth's interior structure.

shearing   The motion of two surfaces sliding past one another.

shield volcano   A volcano built by flows of fluid basaltic lava, in the shape of a dome with gently sloping sides.

silica   A chemical combination of silicon and oxygen.

silicate mineral   A mineral composed mainly of silicon and oxygen.

silicic   A descriptive term for volcanic rock or magma that, like rhyolite, is rich in silica.

skylight   In volcanology, a hole or opening in the crust over an active lava tube.

solfatara   A fumarole whose gases are primarily sulfurous.

spatter cone   A cone built up around a vent by fragments of still-molten lava that weld into a solid mass.

stratosphere   An upper portion of the atmosphere, above about 15 km.

stratovolcano   A steep volcanic cone built by both lava flows and pyroclastic deposits from explosive eruptions.

strike-slip fault   A nearly vertical fault with sideslipping, horizontal displacement.

subduction-type volcano   A volcano that occurs just inland from a subduction zone.

subduction zone   The zone where two tectonic plates converge, usually with one overriding the other.

surge   A transient increase in the velocity and volume of a lava flow.

talus   Rock debris at the base of a steep slope or cliff.

tephra   A term for material of all sizes and types erupted from a volcano and usually deposited by airfall. In a less restricted sense, a synonym for all pyroclastic deposits.

thermal gradient   The rate of change of temperature with depth or distance.

thrust fault   A gently inclined fault whose upper side moves relatively upward.

tidal wave   See *tsunami*.

transform fault   A strike-slip fault connecting the offsets of a mid-ocean ridge.

tremor, harmonic   Volcanic tremor that has a steady frequency and amplitude.

tremor, volcanic   A continuous vibration of the ground, detectable by seismographs, that is associated with volcanic eruptions and other subsurface volcanic activity.

tropopause   The boundary at the base of the stratosphere.

tsunami   A great sea wave produced by a submarine earthquake, landslide, or volcanic eruption.

vein   A mineral deposit precipitated in a rock fracture.

vent   An opening at the Earth's surface through which volcanic materials are erupted.

viscosity   A measure of resistance to flow in a liquid.

volcanic complex   A persistent volcanic vent area that has built a complex mixture of volcanic landforms.

volcanic front   The line of volcanoes closest to an oceanic trench.

wave-cut terrace   A shelf formed by wave erosion of coastal rocks; sometimes uplifted above sea level.

welded tuff   A pyroclastic deposit so hot when formed that the fragments welded together into a solid rock.

yield strength   The stress level that must be exceeded before a plastic material will deform.

Supplementary reading

PART I. VOLCANIC MOUNTAINS

Bullard, Fred M. *Volcanoes of the Earth,* rev. ed. Austin and London: University of Texas Press, 1976.

A thorough but largely descriptive book on volcanic features with many examples of famous eruptions.

Editors of Time–Life Books. *Volcano.* Alexandria, Va.: Time–Life Books, 1982.

Well-written history of volcanology and narratives of some famous eruptions. Excellent color illustrations.

Harris, Stephen. *Fire Mountains of the West.* Missoula, Mont.: Mountain Press, 1987.

Popularly written but accurate information on the volcanoes and volcanic areas of the western United States.

Krafft, Maurice, and Krafft, Katia. *Volcanoes: Earth's Awakening.* Maplewood, N.J.: Hammond, 1980.

An introduction to volcanic activity, beautifully illustrated with color photographs.

McClelland, L., Simkin, T., Summers, M., Neilsen, E., and Stein, T. *Global Volcanism 1975–1985.* Englewood Cliffs, N.J.: Prentice Hall, 1989.

Descriptions of all volcanic eruptions during 1975–85, largely by on-the-scene investigators.

Scientific American. *The Dynamic Earth.* New York: Scientific American, 1983.

Chapters on the Earth's core, mantle, and crust, as well as on the ocean atmosphere, and biosphere by authors expert in those subjects.

Simkin, T., Siebert, L., McClelland, L., Bridge, D., Newhall, C., and Latter, J. H. *Volcanoes of the World.* Washington, D.C.: Smithsonian Institution, 1981.

A catalog of active volcanoes and their eruptive history during the last 10,000 years.

Smithsonian Institution. *Scientific Event Alert Network (SEAN) Bulletin.* Washington, D.C.: Smithsonian Institution, 1977–

A monthly periodical summarizing volcanic, earthquake, and atmospheric fireball events during the preceding month. Names and addresses of sources are also provided.

U.S. Geological Survey. *Earthquakes and Volcanoes.* Reston, Va.: U.S. Geological Survey, 1970–

A bimonthly periodical with timely articles on earthquakes, seismology, volcanoes, and volcanism by authors who are actively involved in scientific research.

Wyllie, Peter J. *The Way the Earth Works.* New York: Wiley, 1976.

Commonsense explanations of plate tectonics and other major Earth processes.

PART II. VOLCANIC ROCKS

Cambridge Encyclopedia of Earth Science, ed. David G. Smith. Cambridge: Cambridge University Press, 1982.

Good sections on volcanic rocks and heat-flow from the Earth's interior.

Cas, R.A.F., and Wright, J. V. *Volcanic Successions.* London: Allen & Unwin, 1987.
Thorough discussion of volcanic products, assemblages of volcanic rocks, and the processes that formed them.

Decker, R. W., Wright, T. L., and Stauffer, P. H., eds. *Volcanism in Hawaii* (Professional Paper 1350). 2 vols. Washington D.C: U.S. Geological Survey, 1987.
Sixty-two chapters covering physiography, tectonics, and submarine geology; geology of the island of Hawaii; petrogenisis and volcanic gases; structure; dynamics; and history of investigations.

Fisher, R. V., and Schmincke, H.-U. *Pyroclastic Rocks.* Berlin: Springer-Verlag, 1984.
Definitive but technical treatise on pyroclastic ejecta from volcanoes.

Francis, Peter. *Volcanoes.* Middlesex, UK: Penguin Books, 1976.
Well written, with emphasis on volcanic products.

Lipman, Peter W., and Mullineaux, Donal R., eds. *The 1980 Eruptions of Mount St. Helens, Washington* (Professional Paper 1250). Washington D.C.: U.S. Geological Survey, 1981.
A collection of 62 reports covering eruptions, geophysical monitoring, volcanic deposits, effects of the 1980 eruptions, and analysis of potential hazards.

PART III. VOLCANIC RISK AND REWARD

Blong, R. J. *Volcanic Hazards.* New York: Academic Press, 1984.
The definitive book on hazards associated with volcanoes, it includes many case studies based on actual eruptions.

Cairns-Smith, A. G. *Seven Clues to the Origin of Life.* Cambridge: Cambridge University Press, 1985.
A short but intriguing book on how clay minerals may have played a major role in the mystery of the origin of life.

Holland, H. D. *The Chemical Evolution of the Atmosphere and Oceans.* Princeton, N.J.: Princeton University Press, 1984.
Definitive but technical treatise on the changing compositions of the Earth's air and oceans.

Rona, Peter A. "Mineral Deposits from Sea-Floor Hot Springs." *Scientific American* 254, no. 1 (January 1986): 84–92.
Well-written and illustrated article explaining mineral deposition on the seafloor and how seafloor spreading can relocate submarine ore deposits on land.

Rybach, L., and Muffler, L.J.P. *Geothermal Systems: Principles and Case Histories,* New York: Wiley, 1981.
Comprehensive explanation and examples of different geothermal systems worldwide.

Stommel, Henry, and Stommel, Elizabeth. *Volcano Weather.* Newport, R.I.: Seven Seas Press, 1983.
The story of 1816, "the year without a summer," and how the severely cold weather might have been caused by the giant eruption of Tambora Volcano in Indonesia.

Index